P.E. Petrides · W.M. Nauseef (Eds.)
The Peroxidase Multigene Family of Enzymes

Springer
Berlin
Heidelberg
New York
Barcelona
Hong Kong
London
Milan
Paris
Singapore
Tokyo

Petro E. Petrides · William M. Nauseef (Eds.)

The Peroxidase Multigene Family of Enzymes

Biochemical Basis and Clinical Applications

With 70 Figures in 100 Parts

Springer

Petro E. Petrides, M.D., Ph.D.
Charité Humboldt University of Berlin, Department of Medicine
Schumannstr. 20/21, 10117 Berlin, Germany

William M. Nauseef, M.D.
Inflammation Program and Departments of Medicine
University of Iowa and Veterans Affairs Medical Center
200 Hawkins Drive SW 54 GH, Iowa City, IA 52242, USA

ISBN 3-540-67154-4 Springer-Verlag Berlin Heidelberg New York

Library of Congress Cataloging-in-Publication Data
The peroxidase multigene family of enzymes : biochemical basis and clinical applications / Petro E. Petrides, William M. Nauseef, eds.
p. ; cm.
Includes bibliographical references and index.
ISBN 3540671544 (hardcover : alk. paper)
1. Peroxidase-Congresses. I. Petrides, Petro E., 1949-II. Nauseef, William M., 1950-
[DNLM: 1. Peroxidases-physiology-Congresses. 2. Disease Susceptibility-enzymology-Congresses. 3. Hereditary Diseases-enzymology-Congresses. 4. Immune System-enzymology-Congresses. 5. Peroxidases-deficiency-Congresses. QU 140 P4527 2000]
QP603.P4 P46 2000
572'.791-dc21   00-028460

Springer-Verlag Berlin Heidelberg New York
a member of BertelsmannSpringer Science + Business Media GmbH

Printed in Germany

Cover Design: design & production GmbH, 69121 Heidelberg
Typesetting: FotoSatz Pfeifer GmbH, D-82166 Gräfelfing
Printed on acid-free paper – SPIN: 10749787   27/3130 – 5 4 3 2 1 0

Preface

First described in the 1940s as verdoperoxidase (because of its green color), the myeloid hemeprotein myeloperoxidase (MPO) was assigned a prominent position in the array of oxygen-dependent microbicidal responses of human neutrophils after Dr. Seymour Klebanoff's seminal publication in 1970 (*Science* 169: 1095, 1970). Over the subsequent decades, investigators from many laboratories have elucidated many of the functional properties of MPO, both as a purified enzyme in vitro and as a granule constituent released within the confines of the neutrophil phagosome. In parallel, investigators have unraveled many of the structural features of MPO, culminating recently in the solution of the crystal structure of MPO at 1.75 Å.

Consistent with the central role of MPO in neutrophil microbicidal activity, reports of inherited MPO deficiency were few and limited to devastating and often fatal systemic infections with *Candida* species. However, in the 1980s the contribution of MPO-dependent events to neutrophil-mediated host defense was reassessed when advances in automated leukocyte enumeration and differentiation were applied clinically. Beginning with the Technicon Hemalog Analyser, and later with more advanced devices developed by Bayer, clinical hematology laboratories utilizing this versatile technology uncovered a previously unrecognized population of MPO-deficient subjects who had escaped medical recognition because of the absence of obvious clinical consequences.

No one made better use of the Bayer analytical system than did Dr. Dolphe Kutter. In his clinical laboratory in Luxembourg he exploited fully the potential of this sensitive and rapid technology to identify the prevalence and spectrum of peroxidase deficiencies, including MPO deficiency and defects affecting monocytes and eosinophils. Dolphe's initiative and Bayer's support prompted a 1-day symposium in October 1996 entitled "Recent Developments in MPO deficiency" and held at the Centre Universitaire in Luxembourg. Invited speakers at that symposium included Dolphe, Dr. Francesco Lanza (Ferrara, Italy), Dr. William M. Nauseef (Iowa City, USA), Dr. Giuseppe d'Onofrio (Rome, Italy), Dr. Petro E. Petrides (Munich, Germany), and Dr. Eric Tschirhardt (Luxembourg). During informal conversations among the speakers and including Wolfgang Kosanke (Bayer), two issues repeatedly surfaced. First, information about the clinical consequences of MPO deficiency, inherited or acquired, was extremely limited. Given the various reports being published at that time implicating MPO in a variety of biological events unrelated to host defense, all agreed that only appropriately designed population-based studies could rigorously test these hypotheses and were sorely needed. Second, it was clear that MPO deficiency represented only a small subset of the biologically important aspects of MPO in particular, and of animal peroxidases in general. Based on these recurring issues, it was decided that another meeting should be convened but with a more broadly inclusive theme. In addition, updated versions of the talks presented at the Luxembourg meeting were published in the September 1998 issue of the Journal of Molecular Medicine (76: 659–698).

To that end, we convened "The Peroxidase Multigene Family of Enzymes: Biochemical and Clinical Applications" from 27 September to 2 October 1998 in the Benedictine Abbey on Fraueninsel in Lake Chiemsee in Upper Bavaria. Scientists from 19 different nations, including Japan, New Zealand, the United States, and many European countries, came to discuss a wide range of subjects related to members of the protein family of animal peroxidases. The serenity of the site facilitated quiet contemplation and the intimate surroundings free of distraction permitted lively interactions and rewarding exchanges of information among the participants. The broad range of topics attracted an accordingly diverse group of scientists, thus affording a unique opportunity for interactions between investigators that seldom attend the same meetings because of their varied scientific orientations. For example, it was possible for clinicians studying the renal consequences of vasculitis associated with antibodies to MPO to discuss their science with structural biologists solving the crystal structure of purified MPO.

As detailed in the papers presented at Chiemsee and published here, the range of topics discussed was extremely broad. New data on the crystal structure of MPO were presented, as were models for the structures of related peroxidases based on that of MPO. Analyses of structure-function relationships of MPO and thyroid peroxidase (TPO), the bases of the microbicidal activity of MPO, the structural consequences of inherited defects in the structures of MPO, TPO, and eosinophil peroxidase (EPO), the prevalence and clinical sequelae of inherited MPO deficiency in Japan and in Europe, and the end-organ morbidity associated with vasculitis related to antibodies to neutrophil granule proteins, including MPO, were all discussed in depth. Provocative findings implicating MPO in the pathobiology of diseases not obviously related to host defense against infection were presented, including diseases such as cystic fibrosis, multiple sclerosis, Alzheimer's disease, and atherosclerosis. Taken together, the topics discussed portrayed an expansive landscape of biology in which the animal peroxidases participated, in some cases in roles not previously recognized. Participants departed Fraueninsel invigorated with a deeper appreciation of the breadth and width of biology, chemistry, clinical medicine, and biochemistry influenced by members of the animal peroxidase protein family. This appreciation has also sparked several international collaborations between attendants of the conference (see also: Peroxidase and human disease: a meeting of minds. Molecular Medicine Today (1999) 5: 58–60).

We hope that reading the contributions of the speakers published in this book will recreate some of the enthusiasm and excitement that was palpable during the meeting. However, even in the short interval since this meeting, significant advances in studies of the animal peroxidases have been made. Murine models of MPO deficiency have been used to test hypotheses concerning the role of MPO in host defense and in inflammatory diseases such as atherosclerosis. The similarities and differences, both structural and functional, among members of the peroxidase protein family continue to accumulate and the extension of the evolutionary "family tree" to seemingly disparate members hints at biological functions for peroxidases that are not currently appreciated. Progress at unraveling the structure-function relationships of the heme group peculiar to MPO continues and promises to provide novel insights into fundamental aspects of the biology of heme proteins in general. Likewise clinicians are advancing our understanding of the pathogenesis of autoimmune diseases associated with antibodies to neutrophil granule proteins including MPO. Given the flurry of productive activity in this area, the next peroxidase meeting promises to surpass in breadth and depth the meeting reported here. We hope readers of this book will consider joining us at Schloss Wilheminenberg in Vienna, Austria, from 3 to 8 September 2000 for "The Peroxidase Superfamily II of Animal and Human Enzymes: biochemical basis and clinical application". Information is available from Drs. Ursula Burner or Christian Obinger in Vienna (www.boku.ac.at/pod2000) and the meeting promises to be a great one.

With regard to the Chiemsee meeting thanks and credits have to be given to several institutions and individuals: Bayer Inc., represented by Wolfgang Kosanke, laid the ground for the financing of the meeting by providing start-up funds, and the Deutsche Forschungsgemeinschaft, Bonn, Germany, contributed – after peer review of the program – a large sum (including travel funds for the attendance of scientists from countries of former Eastern Europe and the Soviet Union) which was supplemented by the Bavarian Ministry for Cultural Affairs, the Charité Berlin as well as University of Iowa. Funds, however, are necessary but not sufficient for the success of such a meeting. Human capital is at least as important: in this regard Sarah Miller of Iowa City and Chuanbing Zang and his wife of Munich and later Berlin contributed many hours and ideas to the preparation of the meeting. They also joined forces at lake Chiemsee with Dr. Maria Beykirch (the wife of one of us) and Sister Scholastika (our warmhearted host at the Abbey) to make people feel comfortable in upper Bavaria.

Last but not least, donations from Bayer Inc., Charité Berlin, Genentech South San Francisco and Kutter Laboratories, Luxembourg, as well as the devoted work of Sven Schmidt and the Springer-Verlag crew and by Dörthe Mennecke-Bühler made the publication of this volume possible.

Petro E. Petrides, Berlin, William M. Nauseef, Iowa City

Contents

1 Peroxidases: A Historical Overview of Milestones in Research on Myeloperoxidase
Robert A. Clark 1

2 Targeting of Proteins to Lysosomes and Granules
Andrej Hasilik 11

3 Sorting and Processing of Neutrophil Granule Proteins
Urban Gullberg, Niklas Bengtsson, Elinor Bülow, Daniel Garwicz, Anders Lindmark, Inge Olsson 22

4 X-Ray Crystallographic Studies of Human Myeloperoxidase
C.A. Davey, T. Fiedler, R.E. Fenna 31

5 Structural and Biological Properties of Human Recombinant Myeloperoxidase
Nicole Moguilevsky, Alex Bollen 38

6 Recent Insights into the Biosynthesis and Processing of Human Myeloperoxidase
William M. Nauseef.................................. 45

7 Targets for Myeloperoxidase-Dependent Bactericidal Activity
Henry Rosen 52

8 Reactions of Myeloperoxidase and Production of Hypochlorous Acid in Neutrophil Phagosomes
Christine C. Winterbourn, Christine J. van Dalen, Mark B. Hampton, Anthony J. Kettle 58

9 Acquisition and Use of Myeloperoxidase in the Microbicidal Activity of Macrophages
Laszlo Marodi, Christopher Tourney, Rita Kaposzta, Richard B. Johnston Jr., Nicole Moguilevsky 68

10 Cooperative Interactions of the Peroxidase and Nitric Oxide Pathways in Inflammatory Oxidant Production
Jason P. Eiserich, Bruce A. Freeman, Carroll E. Cross, Albert van der Vliet 72

11 The Role of the Monoclonal Antibody Anti-Myeloperoxidase (anti-MPO) in the Diagnosis and Classification of Acute Leukaemias
Estella Matutes 86

12 The Role of Myeloperoxidase in Myeloid Leukemia and Multiple Sclerosis
Wanda F. Reynolds 93

13 Pathways for Oxidative Tissue Injury by Myeloperoxidase
Jay W. Heinecke 102

14 Myeloperoxidase Activity of Neutrophils in Cystic Fibrosis
Véronique Witko-Sarsat 107

15 Structure Function Relationships Amongst Members of the Animal Peroxidase Family of Proteins
Rosa Pia Ferrari, Silvio Traversa 114

16 Myeloperoxidase (MPO) and Eosinophil Peroxidase (EPO) Deficiency in the Region Friuli-Venezia Giulia of North-Eastern Italy
Pierluigi Patriarca, Maurizio Romano 122

17 Biochemistry and Physiology of Thyroid Peroxidase
Jan J.M. de Vijlder, Hennie Bikker 129

18 Severe Congenital Hypothyroidism Caused by Mutations in the Thyroid Peroxidase Gene
Hennie Bikker, Jan J.M. de Vijlder 133

19 Screening for Leukocyte Peroxidase Deficiencies by Means of Flow Cytometry: Application to the Study of Prevalence, Pathology and Genetics
Dolphe Kutter, Luc Verstraeten 136

20 Prevalence of Inherited Myeloperoxidase Deficiency in Japan
Kazuo Suzuki, Hiroyuki Nunoi, Makoto Miyazaki, Fumikazu Koi 145

21 Acquired and Inherited Forms of Myeloperoxidase Deficiency: Clinical and Hematological Features
Francesco Lanza, Angela Latorraca, Sabrina Moretti, Barbara Castagnari, Luisa Ferrari, Gianluigi Castoldi 150

22 Myeloperoxidase-Positive and Negative Granulocytes: Investigations by Flow Cytometry
Gernot Bruchelt, Claudia E. Gerber, Peter Bader, Matthias Zipfel, Thomas Klingebiel, Rupert Handgretinger, Dietrich Niethammer 157

23 Mutation Analysis for Genotype-Phenotype Relationships in Myeloperoxidase Deficiency
Petro E. Petrides, Susanne Bock, Chuanbing Zang 166

24 Diagnostic Significance of Antineutrophil Cytoplasmic Antibodies (ANCA) in Systemic Vasculitides
Loïc Guillevin, Bernard Jarrousse 173

25 Immunodiagnostic Aspects of Autoantibodies Against Myeloperoxidase
Elena Csernok, Wolfgang L. Gross 180

26 Pathophysiological Mechanisms in Anti-Myeloperoxidase-Associated Vasculitis
Jan Willem Cohen Tervaert 186

Subject Index 191

Participants of the peroxidase meeting

Contributors

H. Bikker, Academic Medical Center, University of Amsterdam, Emma Children's Hospital, Pediatric Endocrinology, Meibergdreef 9, 1105 AZ Amsterdam, The Netherlands

G. Bruchelt, Children's University Hospital, Department of Hematology and Oncology, Hoppe-Seyler-Str.1, 72076 Tuebingen, Germany

R. Clark, Department of Medicine, University of Texas Health Science Center and South Texas Veterans Health Care System, San Antonio, TX 78284, USA

E. Csernok, Department of Rheumatology, Medical University of Lübeck and Rheumaklinik, Oskar-Alexander-Str. 26, 24576 Bad Bramstedt, Germany

C. Davey, Institut für Molekularbiologie und Biophysik, ETH Hoenggerberg HPM, 8093 Zürich, Switzerland

J.P. Eiserich, Department of Anesthesiology and the UAB Center for Free Radical Biology, University of Alabama at Birmingham, Birmingham, 619 19th Street South, 946 Tinsley Harrison Tower, AL 35233, USA

R.P. Ferrari, Dipartimento di Chimica IFM, Università di Torino, via P. Giuria 7, 10125 Torino, Italy

L. Guillevin, Department of Internal Medicine, Hôpital Avicenne, University of Paris, 125 rue de Stalingrad, 93000 Bobigny, France

U. Gullberg, Department of Hematology, Research Department 2, E-block, University Hospital, S-221 85 Lund, Sweden

A. Hasilik, Philipps-Universität Marburg, Institute of Physiological Chemistry, Karl-von-Frisch-Str. 1, 35033 Marburg, Germany

Jay W. Heinecke, Department of Medicine and Department of Molecular Biology and Pharmacology, Washington University School of Medicine, St. Louis, MO 63110, USA

D. Kutter, Laboratoires réunis, PO Box 11, 6101 Junglinster, Grand-Duchy of Luxembourg

F. Lanza, Hematology Section, Department of Biomedical Sciences and Advanced Therapies, University of Ferrara, St Anna Hospital, Corso Giovecca No. 203, 44100 Ferrara, Italy

L. Marodi, Department of Pediatric Infectious Diseases and Immunology University School of Medicine, Nagyerdei Krt. 98, 4012 Debrecen, Hungary

E. Matutes, Academic Department of Haematology and Cytogenetics.The Royal Marsden Hospital and Institute of Cancer Research, Fulham Road. London SW3 6JJ, UK

N. Moguilevsky, Applied Genetics, Université Libre de Bruxelles, rue de l'Industrie 24, B-1420 Nivelles
Present address: Institut de Biologie et de Médecine Moleculaires, Service de Génétique Appliquée, rue des professeurs Jeener et Brachet 12, B-6041 Gosselies

W.M. Nauseef, Inflammation Program and Departments of Medicine, University of Iowa and VA Medical Center, 200 Hawkins Dr. SW 54 GH, Iowa City, IA 52242, USA

P. Patriarca, Department of Physiology and Pathology, School of Medicine, University of Trieste, Via A. Fleming 22, 34127 Trieste, Italy

P.E. Petrides, Department of Medicine, Charité Humboldt University of Berlin, Klinik und Poliklinik II mit Schwerpunkt Hämatologie und Onkologie, Campus Mitte Charité, Humboldt-Universität zu Berlin, Schumannstr. 20/21, 10117 Berlin,
Present adress: Hematology Oncology Center, Zweibrückenstr. 2, 80331 Munich, Germany

W.F. Reynolds, Sidney Kimmel Cancer Center, 10835 Altman Row, San Diego, CA 92121, USA

H. Rosen, Department of Medicine, University of Washington, Box 356420, Seattle, WA 98195, USA

K. Suzuki, Bayer-Sankyo Co. Ltd.,Takamatsu Hospital, and Department of Bioactive Molecules, National Institute of Infectious Diseases, 23-1 Toyama 1-Chome, Shinjuku-ku, Tokyo 162-8640, Japan

J.W.C. Tervaert, Department of Clinical Immunology, University Hospital, Groningen, Hanzeplein 1, 9713 GZ Groningen
Present adress: Academisch Ziekenhuis Maastricht, Afdeling Klinische Immunologie, Postbus 5800, 6202 AZ Maastricht, The Netherlands

J.J.M. de Vijlder, Academic Medical Center, University of Amsterdam, Emma Children's Hospital, Pediatric Endocrinology, Meibergdreef 9, 1105 AZ Amsterdam, The Netherlands

C.C. Winterbourn, Free Radical Research Group, Department of Pathology, Christchurch School of Medicine, PO Box 4345, Christchurch, New Zealand

V. Witko-Sarsat, INSERM U 507, Hôpital Necker Enfants Malades, 75015 Paris, France

Chapter 1

Peroxidases: A Historical Overview of Milestones in Research on Myeloperoxidase

Robert A. Clark

1.1 Introduction

Peroxidases are a group of enzymes that catalyze the oxidation of a variety of electron donors by hydrogen peroxide (H_2O_2). Among the members of the group discussed at this meeting are myeloperoxidase (MPO), eosinophil peroxidase (EPO), lactoperoxidase (LPO), and thyroid peroxidase (TPO). Although there are significant structural and functional similarities among members of the peroxidase family (see Chap. 15), each of these peroxidases has a highly specific and very restricted range of tissue expression (Table 1.1). This historical review focuses predominantly on MPO as a representative member of this family of enzymes and perhaps the most thoroughly studied of the animal peroxidases [1]. MPO is present in the neutrophil (or heterophil) at very high concentrations, making up as much as 5% of the dry weight of the cell. Lesser amounts are found in blood monocytes, but tissue macrophages are generally lacking MPO. The enzyme has been detected in the leukocytes of a wide range of vertebrate species and of those studied is absent only from chicken and goose.

Table 1.1. Selected peroxidases of animal tissue

Peroxidase	Tissue expression
Myeloperoxidase	Neutrophils, mononuclear phagocytes
Eosinophil peroxidase	Eosinophils
Lactoperoxidase	Milk, saliva, other secretions
Thyroid peroxidase	Thyroid

1.2 Early Studies

The study of peroxidases dates to the middle of the nineteenth century when a number of investigators began examining both plant and animal tissues using various organic compounds whose color properties varied with oxidation state. Selected milestones from that period through the mid-twentieth century are summarized briefly in Table 1.2 (see reference 2 for all original citations). It was Schönbein who first reported in 1855 that various plant and animal tissues oxidize guaiac, implying the presence of peroxidase activity. The first detection of guaiac oxidation by pus, presumably representing MPO activity, was reported by Klebs in 1868, and then by Struve several years later. In the 1880s Felix Hoppe-Seyler in Tübingen applied these

Table 1.2. Peroxidases – historical background (see Agner [2] for all original citations)

1855 Schönbein	Guaiac oxidation by plant and animal tissues
1868 Klebs; 1872 Struve	Guaiac oxidation by pus[a]
1883 Hoppe-Seyler	"Activation" of oxygen by tissues
1885 Ehrlich	Indophenol blue reactions of living tissues ("oxygen need") "der ungesättigte Zustand des Protoplasmas"
1895 Röhmann and Spitzer	Indophenol blue oxidation by tissue ferments
1897 Spitzer	Iron bound to the oxygen activator
1898 Linossier	Hydrogen peroxide required for leukocyte oxidation reactions[a]
1899 Raudnitz; 1901 Loew	Two distinct activities of tissue ferments: catalases and peroxidases
1907 Winkler	Histochemistry of peroxidase in leukocyte cytoplasmic grains[a]
1910 Fischel	Peroxidase in cytoplasmic extracts of neutrophils and eosinophils[a]
1922 Graff	Inhibition of peroxidases by cyanide (cf. cytochromes of Warburg and Keilin)
1924 Neumann	Partially purified peroxidase from horse eosinophils and neutrophils[a]
1941 Agner	Purification of "verdoperoxidase" from human leukocytes[a]

[a] Studies on leukocytes

techniques further and developed the concept of "activation" of oxygen by living tissues. Also at this time Paul Ehrlich carried out detailed studies of living tissues using the indicator indophenol blue and advanced the concept of "oxygen need" as a basis for the varied oxidation-reduction reactions of different tissues. Röhmann and Spitzer demonstrated similar dye oxidation reactions by soluble tissue extracts designated ferments.

Advances in understanding the chemistry of biological oxidative reactions occurred when Spitzer demonstrated in 1897 that the oxygen activator in tissue ferments contained bound iron and Linossier reported in 1898 the key observation that H_2O_2 is required for oxidation reactions carried out by white blood cells. At about the turn of the 19th century Raudnitz and Loew independently established the important distinction between two different activities of tissue ferments, namely catalases and peroxidases. That blood leukocytes contain peroxidase activity was shown histochemically by Winkler in 1907. This demonstration of peroxidase-positive leukocyte cytoplasmic "grains" was an early forerunner of modern-day understanding of the various classes of granular organelles found in the cytoplasm of leukocytes. Also at about that time Fischel demonstrated the presence of peroxidase activity in cytoplasmic extracts of neutrophils and eosinophils. In 1922, Graff found that peroxidases are inhibited by cyanide, thus establishing a parallel to the ongoing work on cytochromes by such investigators as Warburg and Keilin. The first partial purification of leukocyte peroxidase was achieved by Neumann in 1924 using a mixture of horse eosinophils and neutrophils.

1.3 Purification

A major milestone in research on MPO occurred in 1941 when Kjell Agner [2] reported the purification of an enzymatically active protein from human leukocytes (Fig. 1.1). The bright-green color of the purified product led Agner to designate it as "verdoperoxidase", although later realization that other peroxidases were green in color resulted in a modification of the terminology and the leukocyte peroxidase has

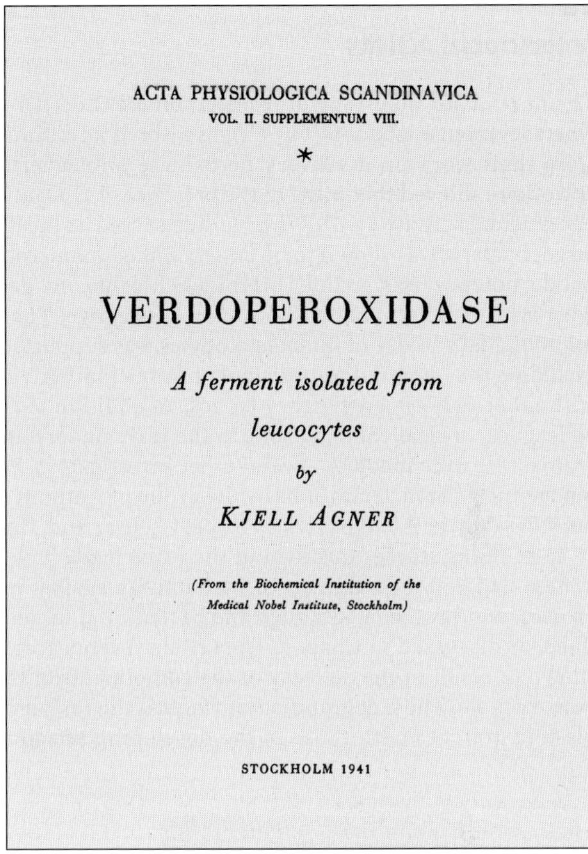
ACTA PHYSIOLOGICA SCANDINAVICA
VOL. II. SUPPLEMENTUM VIII.

*

VERDOPEROXIDASE

A ferment isolated from leucocytes

by

KJELL AGNER

(From the Biochemical Institution of the Medical Nobel Institute, Stockholm)

STOCKHOLM 1941

Fig. 1.1. Title page from 1941 publication by Agner [2]

been known for some years as MPO. Interestingly, Agner had started out with the intention of purifying catalase rather than peroxidase. His earlier attempts to purify catalase from liver tissue were complicated by contamination with various liver pigments. Searching for an alternative tissue source, he chose the leukocyte-rich empyema fluid from patients with tuberculous pleural infection, a common disease at the time. The major protein isolated had all of the spectral, physical, biochemical, and enzymatic properties now known to be characteristic of MPO. In addition, Agner isolated two other pigmented proteins, one of which had the properties of a b-type cytochrome. It is tempting to speculate that this may have represented a very early and preliminary characterization of flavocytochrome b_{558}, the cytochrome now known to be a key component of the neutrophil respiratory burst oxidase. In a subsequent paper in 1958, Agner reported a new method for MPO purification using infected dog uteri as a tissue source and an improved adsorption chromatographic procedure [3] that served as the standard method for purification of the enzyme for some time.

1.4 Antimicrobial Activity

The next major milestone in research on MPO occurred in the middle to late 1960s when Seymour Klebanoff (Fig. 1.2) discovered its antimicrobial properties. Extending from their work on a salivary peroxidase antibacterial system [4], Klebanoff and coworkers showed that MPO in the presence of H_2O_2 and a halide exhibit very potent bactericidal activity [5, 6]. When iodide served as the halide, the direct iodination of target bacteria was shown to be closely linked to microbial killing [7, 8]. The order of halide potency was iodide>bromide>chloride. In general, fluoride was inactive, whereas the pseudohalide thiocyanate was active. That MPO was important for the microbicidal activity of intact leukocytes was supported by several lines of evidence, including the impaired bactericidal or fungicidal activity of leukocytes from patients with inherited MPO deficiency [9, 10]. In addition to the work of Klebanoff and his colleagues, critical contributions to the early understanding of MPO and its antimicrobial role were made by several other investigators. Perhaps most notable were the biochemical characterizations by the group of Anthony Sbarra [11, 12] and the studies on MPO-deficient leukocytes by Robert Lehrer and Martin Cline [9, 13].

Continuing studies throughout the 1970s made it clear that the MPO-H_2O_2-halide system had very broad microbicidal activity against many types of bacteria, fungi, viruses, and protozoa, although the bactericidal targets have not been precisely defined, as discussed in Chap. 7. The cellular events accompanying phagocytosis were shown to result in the delivery of the components of the MPO system to the phagocytic vacuole. Thus, degranulation features the fusion of the membranes of the cytoplasmic granules with those of the developing phagocytic vacuole, resulting in the

Fig. 1.2. Recent photograph of Seymour J. Klebanoff, M.D., Ph.D

delivery of granular contents such as MPO into the phagolysosome. Activation of the respiratory burst oxidase, now known to comprise a multicomponent activatable NADPH oxidase [14], results in the delivery of the products of oxygen reduction to the phagocytic vacuole where a variety of events contribute to the highly toxic milieu (see Chap. 8). The initial product of the oxidase is the superoxide anion ($O_2\cdot^-$), which through a dismutation reaction forms H_2O_2 for utilization by MPO. In general, the most relevant halide for the MPO system is chloride, which is fully active at physiologic concentrations.

1.5 Biochemical Basis of Toxicity

Figure 1.3 illustrates the main biochemical pathways through which MPO mediates its effects. H_2O_2 at relatively low concentrations reacts with the heme iron of MPO to form compound I, a primary catalytic complex capable of oxidizing a variety of substrates, with the regeneration of native MPO [15]. When chloride is the substrate the initial product is hypochlorous acid (HOCl). When the concentration of H_2O_2 is substantially in excess of the concentration of MPO, compound I is converted to the inactive compound II. Various reducing agents, including the superoxide anion can reduce compound II back to active MPO. Another complex, compound III, an oxyperoxidase, can be formed by reactions with various agents, including superoxide. However, the latter reaction is slow relative to that of H_2O_2 with MPO to form compound I. It is clear that the products of the neutrophil respiratory burst, namely superoxide anion and H_2O_2, play multiple roles within the phagocytic vacuole in terms of their reactivity with either native MPO or the subsequent complexes formed (see Chap. 8).

As illustrated in Fig 1.4, the toxicity of the MPO system appears to be mediated through several pathways and a rather wide range of reactive products [14]. When a halide is oxidized by MPO and H_2O_2 the initial product is the hypohalous acid, for example HOCl. With a pKa of 7.53, HOCl exists as a mixture of the nondissociated acid and the hypochlorite ion (OCl^-) at near-neutral pH. Under the acidic conditions within the phagocytic vacuole the predominant form is HOCl, which may react with excess chloride to form chlorine (Cl_2). Thus, the initial products of chloride oxidation (HOCl, OCl^-, and Cl_2) comprise a group of powerful oxidants. However, further toxicity is achieved through additional reactions, including the chlorination of tyrosine

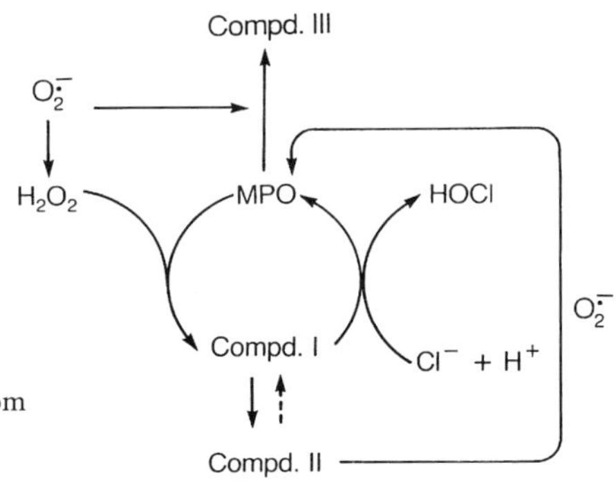

Fig. 1.3. Reaction pathways of MPO, H_2O_2, and superoxide anion. Reproduced with permission from an original illustration by Seymour J. Klebanoff, University of Washington, Seattle, Washington

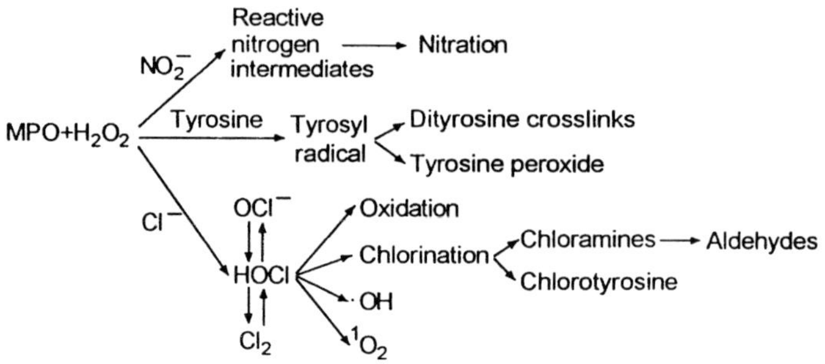

Fig. 1.4. Toxic products generated through the reactions of MPO and H_2O_2 with chloride, tyrosine, and nitrite. Reproduced with permission from an original illustration by Seymour J. Klebanoff, University of Washington, Seattle, Washington

to form chlorotyrosine or reaction with nitrogen centers to form mono- and dichloramines. In the case of amino acid chloramines, oxidative deamination and decarboxylation result in the formation of the corresponding aldehydes. HOCl may serve as an intermediate in the formation of hydroxyl radical (·OH) through its reaction with superoxide. Although the use of EPR spectroscopy along with chemical spin traps has provided evidence for this reaction pathway in leukocytes, the utilization of superoxide for hydroxyl radical formation appears to be quantitatively limited. The reaction of hypochlorite with hydrogen peroxide can lead to the formation of singlet molecular oxygen ($^1O_2$), although whether this reaction takes place to a significant degree in neutrophils remains controversial.

Under certain experimental conditions MPO and H_2O_2 are capable of oxidizing substances other than halides, for example nitrite and tyrosine (Fig. 1.4). MPO-catalyzed oxidation of nitrite results in the nitration of tyrosine, either as the free amino acid or as a residue within proteins. In addition, nitrating compounds are formed through the interaction of nitrite with HOCl. Whether such reactions increase or decrease overall toxicity depends upon the relative potency of HOCl versus the products of its reaction with nitrite. MPO and H_2O_2 are capable of reacting directly with tyrosine in either its free or protein-bound form. The tyrosyl radical thus formed can crosslink generating dityrosine linkages or react with superoxide to form tyrosine peroxide. Even from this incomplete summary of reaction pathways, it is clear that the toxicity of the MPO system is a complex function of multiple potential products. The breadth of physiologic reactions involving MPO-dependent systems are discussed in greater detail by Drs. Eiserich (Chap. 10) and Heinecke (Chap. 13).

1.6 Extracellular Activity and Role in Tissue Injury

The primary site of the microbicidal activity of the MPO system is within the phagocytic vacuole. However, a number of lines of evidence clearly indicate that the components of the system are secreted to the extracellular fluid and that the products of MPO and H_2O_2 exert toxic activity in these sites [16].

Extracellular antimicrobial activity is expressed against large non-ingestible microbes, including parasites such as *Schistosoma* and *Trichinella* and the hyphal forms of pathogenic fungi. Adjacent host cells are also subject to injury by the secreted products of the peroxidase system and susceptible target cells include tumor cells, erythrocytes, leukocytes, platelets, fibroblasts, and spermatozoa. Products of MPO and H_2O_2 can also react with a wide range of soluble molecules. Attention has been directed toward the oxidative inactivation of a number of biological mediators, including leukocyte chemoattractants, protease inhibitors, microbial toxins, leukocytic enzymes, leukotrienes and iron-binding proteins. In some instances a specific chemical mechanism has been implicated, such as the oxidation of critical methionine residues in the cases of certain chemoattractants and protease inhibitors [16].

Given the broad reactivity and toxicity of the products of the MPO system as well as their extracellular and intracellular sites of action, it is not surprising that a number of different biological roles for the peroxidase system have been proposed and, in fact, demonstrated. Thus, MPO plays significant roles in antimicrobial host defenses, modulation of inflammatory reactions, and tissue injury associated with inflammation. Experimental examples of MPO-mediated injury in vivo include pulmonary epithelial injury and fibrosis [17], certain forms of glomerulonephritis [18, 19], and atherosclerosis [20], features of which are presented by Drs. Eiserich (Chap. 10), Heinecke (Chap. 13), Guillevin (Chap. 24), Csernok (Chap. 25), and Tervaert (Chap. 26).

1.7 MPO Deficiency

Studies of patients with hereditary MPO deficiency have contributed substantially to our understanding of the functions of the peroxidase system. Initially it appeared that this genetic disorder was exceedingly rare and the few cases described were seen in patients with histories of infection, especially disseminated candidiasis [9, 13]. In vitro studies showed a correlation with this clinical finding in that the microbicidal defect in MPO-deficient neutrophils was more pronounced for *Candida* than for bacterial species. However, in the early 1980s it became apparent that hereditary MPO deficiency was actually a rather common genetic disorder and that most patients remained healthy in spite of the absence of MPO from their leukocytes. This discovery was based on the introduction of new technology for carrying out automated leukocyte differential counting using a peroxidase staining method [21] as discussed in greater detail by Dr. Kutter (Chap. 19). Thus, it was shown that partial or complete MPO deficiency has a prevalence of approximately 1 per 2000 individuals in the USA and Europe (Chaps. 16 and 19) but occurs much less frequently in Japan (Chap. 20).

Yet to be fully explained is the relatively good health of most MPO-deficient patients compared with the severe phenotype of recurring and life-threatening infections in patients whose leukocytes fail to produce H_2O_2, as is seen in the hereditary disorder chronic granulomatous disease (CGD). In vitro studies have shown that the microbicidal defect in MPO-deficient cells is less severe than that in CGD cells and in general a pattern of delayed rather than absent killing is observed in MPO deficiency [10]. In the absence of MPO the post-phagocytic respiratory burst is more sustained than in normal neutrophils [22, 23]. Therefore, the enhanced production of superoxide, H_2O_2, and other products of oxygen metabolism may explain the relatively limited antimicrobial defect in MPO-deficient cells, in spite of the proven role of MPO in the killing activity of normal cells. In addition, MPO-deficient cells have fully intact oxygen-independent antimicrobial systems, such as defensins, lysozyme, and other cationic proteins. As discussed by Dr. Lanza (Chap. 21), the prevalence of malignancy may be higher in MPO-deficient subjects, although additional data from prospective studies are needed to substantiate this suspected association.

1.8 Molecular Analysis

Research in recent years has taken advantage of the tools of cellular and molecular biology to define the structure, function, and regulation of MPO. In 1987, four different groups independently reported the molecular cloning and characterization of the cDNA for human MPO [24–27]. The group of Morishita then followed up the cDNA cloning with a characterization of the human MPO gene [28], which comprises 12 exons and 11 introns occupying approximately 10 kb on chromosome 17.

The primary translation product of the human MPO gene undergoes extensive processing [29, 30] to yield the glycosylated, heme-containing, mature protein comprising a disulfide-linked 150-kDa dimer of two heavy-chain–light-chain protomers. Two endoplasmic reticulum molecular chaperones, calreticulin and calnexin, are involved in the maturation of the nascent protein [31, 32]. Several specific genetic mutations have been identified in families with inherited MPO deficiency and most individuals with complete deficiency exhibit compound heterozygosity for two different mutations. Detailed analyses of two such MPO gene mutations by Nauseef and coworkers [33, 34] have demonstrated aberrant post-translational processing at two different stages in protein maturation. As highlighted by Dr. Petrides (Chap. 23), complete understanding of the pattern of inheritance in MPO deficiency requires a more complete elucidation of its normal biosynthesis and intracellular trafficking. Targeting of MPO to the azurophilic granule has been the subject of considerable study, and data presented by Drs. Hasilik (Chap. 2) and Gullberg (Chap. 3) demonstrate the potentially important role of the pro-region of MPO in dictating the fate of the mature protein.

1.9 Summary

Based on the foregoing descriptions, the history of research on MPO can be divided into four segments. During the period from the mid-nineteenth century to nearly the mid-twentieth century, tissue oxidative and then peroxidative activities were detected and some features of the leukocyte peroxidase were gradually characterized. The second phase began with Agner's purification of MPO in 1941. This method and subsequent refinements allowed more detailed biochemical studies of MPO and led to the third phase of MPO research beginning in the mid-1960s. Ushered in by the work of Klebanoff, this period extended for approximately 20 years during which the biochemical and biological properties of MPO were characterized in detail, including especially its role in antimicrobial host defenses and later its function in modulating inflammation and mediating host tissue injury. From the mid 1980s and continuing through the present time, detailed cellular, molecular, and genetic analyses of MPO have been carried out. Although a great deal is known, many challenges remain before a full understanding of the biology of MPO and the other animal peroxidases is achieved.

Acknowledgements. The author's research in this area has been supported by the United States Public Health Service (National Institutes of Health grant no. AI 20866) and the Medical Research Service of the United States Department of Veterans Affairs.

References

1. Klebanoff SJ, Clark RA (1978) The neutrophil: function and clinical disorders. Elsevier North Holland, Amsterdam
2. Agner K (1941) Verdoperoxidase: a ferment isolated from leucocytes. Acta Physiol Scand 2 [Suppl 8]: 1–62
3. Agner K (1958) Crystalline myeloperoxidase. Acta Chem Scand 12: 89–94
4. Klebanoff SJ, Luebke RG (1965) The antilactobacillus system of saliva. Role of salivary peroxidase. Proc Soc Exp Biol Med 118: 483–486
5. Klebanoff SJ, Clem WH, Luebke RG (1966) The peroxidase-thiocyanate-hydrogen peroxide antimicrobial system. Biochim Biophys Acta 117: 63–72
6. Klebanoff SJ (1968) The myeloperoxidase-halide-hydrogen peroxide antibacterial system. J Bacteriol 95: 2131–2138
7. Klebanoff SJ (1967) Iodination of bacteria: a bactericidal mechanism. J Exp Med 126: 1063–1078
8. Klebanoff SJ, White LR (1969) Iodination defect in the leukocytes of a patient with chronic granulomatous disease of childhood. N Engl J Med 280: 460–466
9. Lehrer RI, Cline MJ (1969) Leukocyte myeloperoxidase deficiency and disseminated candidiasis: the role of myeloperoxidase in resistance to Candida infection. J Clin Invest 48: 1478–1488
10. Klebanoff SJ (1970) Myeloperoxidase: contribution to the microbicidal activity of intact leukocytes. Science 169: 1095–1097
11. McRlipley RJ, Sbarra AJ (1967) The role of the phagocyte in host-parasite interactions. XII. Hydrogen peroxide-myeloperoxidase bactericidal system in the phagocyte. J Bacteriol 94: 1425–1430
12. Selvaraj RJ, Paul BB, Strauss RR, Jacobs AA, Sbarra AJ (1974) Oxidative peptide cleavage and decarboxylation by the MPO-H_2O_2-Cl^- antimicrobial system. Infect Immun 9: 255–260
13. Salmon SE, Cline MJ, Schultz J, Lehrer R (1970) Myeloperoxidase deficiency. Immunologic study of a genetic leukocyte defect. N Engl J Med 282: 250–253
14. DeLeo FR, Quinn MT (1996) Assembly of the phagocyte NADPH oxidase: molecular interaction of oxidase proteins. J Leukoc Biol 60: 677–691
15. Klebanoff SJ (1999) Myeloperoxidase. Proc Assoc Am Phys 111: (383-389)
16. Clark RA (1983) Extracellular effects of the myeloperoxidase-hydrogen peroxide-halide system. In: Weissmann G (ed) Advances in inflammation research. Raven Press, New York, pp 107–146
17. Johnson KJ, Fantone JC, Kaplan J, Ward PA (1981) In vivo damage of rat lungs by oxygen metabolites. J Clin Invest 67: 983–993
18. Johnson RJ, Couser WG, Chi EY, Adler S, Klebanoff SJ (1987) New mechanism for glomerular injury. Myeloperoxidase-hydrogen peroxide-halide system. J Clin Invest 79: 1379–1387
19. Johnson RJ, Klebanoff SJ, Ochi RF, Adler S, Baker P, Sparks L, Couser WG (1987) Participation of the myeloperoxidase-H_2O_2-halide system in immune complex nephritis. Kidney Int 32: 342–349
20. Hazell LJ, Arnold L, Flowers D, Waeg G, Malle E, Stocker R (1996) Presence of hypochlorite-modified proteins in human atherosclerotic lesions. J Clin Invest 97: 1535–1544
21. Parry MF, Root RK, Metcalf JA, Delaney KK, Kaplow LS, Richar WJ (1981) Myeloperoxidase deficiency. Prevalence and clinical significance. Ann Intern Med 95: 293–301
22. Rosen H, Klebanoff SJ (1976) Chemiluminescence and superoxide production by myeloperoxidase-deficient leukocytes. J Clin Invest 58: 50–60
23. Nauseef WM, Metcalf JA, Root RK (1983) Role of myeloperoxidase in the respiratory burst of human neutrophils. Blood 61: 483–491
24. Johnson KR, Nauseef WM, Care A, Wheelock MJ, Shane S, Hudson S, Koeffler HP, Selsted M, Miller C, Rovera G (1987) Characterization of cDNA clones for human myeloperoxidase: predicted amino acid sequence and evidence for multiple mRNA species. Nucleic Acids Res 15: 2013–2028
25. Morishita K, Kubota N, Asano S, Kaziro Y, Nagata S (1987) Molecular cloning and characterization of cDNA for human myeloperoxidase. J Biol Chem 262: 3844–3851
26. Weil SC, Rosner GL, Reid MS, Chisholm RL, Farber NM, Spitznagel JK, Swanson MS (1987) cDNA cloning of human myeloperoxidase: decrease in myeloperoxidase mRNA upon induction of HL-60 cells. Proc Natl Acad Sci U S A 84: 2057–2061

27. Yamada M, Hur S-J, Hashinaka K, Tsuneoka K, Saeki T, Nishio C, Sakiyama F, Tsunasawa S (1987) Isolation and characterization of a cDNA coding for human myeloperoxidase. Arch Biochem Biophys 255: 147–155
28. Morishita K, Tsuchiya M, Asano S, Kaziro Y, Nagata S (1987) Chromosomal gene structure of human myeloperoxidase and regulation of its expression by granulocyte colony-stimulating factor. J Biol Chem 262: 15208–15213
29. Nauseef WM (1986) Myeloperoxidase biosynthesis by a human promyelocytic leukemia cell line: insight into myeloperoxidase deficiency. Blood 67: 865–872
30. Nauseef WM (1987) Posttranslational processing of a human myeloid lysosomal protein, myeloperoxidase. Blood 70: 1143–1150
31. Nauseef WM, McCormick SJ, Clark RA (1995) Calreticulin functions as a molecular chaperone in the biosynthesis of myeloperoxidase. J Biol Chem 270: 4741–4747
32. Nauseef WM, McCormick SJ, Goedken M (1998) Coordinated participation of calreticulin and calnexin in the biosynthesis of myeloperoxidase. J Biol Chem 273: 7107–7111
33 Nauseef WM, Cogley M, McCormick S (1996) Effect of the R569W missense mutation on the biosynthesis of myeloperoxidase. J Biol Chem 271: 9546–9549
34. DeLeo FR, Goedken M, McCormick SJ, Nauseef WM (1998) A novel form of hereditary myeloperoxidase deficiency linked to endoplasmic reticulum proteasome degradation. J Clin Invest 101: 2900–2909

Chapter 2

Targeting of Proteins to Lysosomes and Granules

Andrej Hasilik

2.1 Introduction

This chapter is intended to categorize and briefly explain delivery pathways to lysosomes and granules in mammalian cells and vacuoles in yeast cells while highlighting a few paradigms. It reflects recent developments in this area of research and covers only a fragment of the vast literature on the biogenesis of the lysosomal/vacuolar apparatus and the specialized sorting principles. The author apologizes to those whose contributions could not be included in this highly focused and brief treatise outlining different vesicular and non-vesicular routes of lysosomal targeting. Because of limited space, secondary literature and reviews are frequently cited and relevant literature on neurovesicles is omitted.

2.2 Vesicular Transport

The majority of lysosomal proteins are probably synthesized in the endoplasmic reticulum (ER). These might be delivered to lysosomes via a direct vesicular pathway. However, in most if not all cases the proteins appear to pass a series of intermediary compartments such as the Golgi apparatus. Thus they are repeatedly packaged into budding vesicles with characteristic coats and delivered by specific docking and fusion into the next compartment, eventually reaching a lysosome or a late endosome that may have the potential to mature into a lysosome. The packaging and targeting of the vesicles seems to utilize general principles that are exemplified by members of a few families of proteins, as recently reviewed [1]. It is amazing that a few soluble proteins are sufficient for the formation of coated vesicles from artificial lipid membranes. However, budding from natural membranes appears to be facilitated by integral membrane proteins (F. Wieland, personal communication).

The features of the artificial system are well illustrated by formation of COPII-coated vesicles in a system reconstituted from purified yeast proteins and liposomes containing phosphatidyl inositol (PI) 4-phosphate or PI 4,5-bisphosphate [2]. COPII is a coatomer that participates in antegrade transport from the ER to the Golgi. Among these proteins sufficient to mediate coated-vesicle formation is Sar1p, a low molecular weight G-protein that was loaded with a non-hydrolyzable GTP analogue for use in the in vitro reconstitution studies [2]. In vivo the responsible activator and coatomer proteins interact with each other and with the cytosolic tails of membrane proteins of the donor organelle, including receptors that mediate the selection of given cargo molecules. In addition, other proteins specify the compartments for budding and are responsible for the localized activation of the small G-proteins. In the case of COPII-coated vesicles, Sec12p, an integral membrane protein of the yeast ER, serves as a nucleotide exchange factor.

In human cells similar guanosine nucleotide exchange factors, when bound to GTP, are localized to various donor organelles, whereas their GDP-bound forms comprise a soluble pool in the cytosol. They belong to the Sec7 family of the exchange factors and are referred to as adenosyl ribosylation factor (Arf) nucleotide site openers, such as ARNO3 that cooperates with Arf1 in the ER [3]. These Arfs are low molecular weight G-proteins of the *ras* family and their structure and interacting site has been elucidated at 2.2 Å resolution [4]. Currently, six Arfs are known in mammalian cells and two in yeast [5]. Among these, Arf1 and Arf6 are involved in budding from Golgi and plasma membranes, respectively. These G-proteins bear myristoyl residues that are available for membrane insertion when activated with GTP. The myristoyl moiety supports the localized cooperation of Arf with other proteins at the budding site. The GTPase activity of Arf1 depends on both a GTPase activating protein (GAP), that binds to the "switch region 2" of Arf1, and the coatomer. The latter can increase the activity of the Arf1 GTPase by several orders of magnitude and is likely to do so by complementing the active site of the GTPase [6].

In an artificial system, similar to events mediated by COPII, budding of COPI vesicles also can be induced in the absence of integral membrane proteins [7]. The COPI coat protein complex participates in both the antegrade and retrograde transport pathways between the ER and the Golgi and within the Golgi. The selection of the retrogradely transported molecules involves retrieval signals such as KKXX and KXKXX in cytosolic tails of cargo receptors [8]. In vitro formation of COPI-coated vesicles from phospholipid vesicles requires only the presence of coatomer, Arf1, and GTP or its non-hydrolyzable analogues. Arf is needed as a partner in a structural interaction rather than as a catalyst [9]. The phospholipid requirements of COPI and COPII vesicle budding differ and the distinct contents of these lipids in different organelles may facilitate the localization of COPI to the Golgi and endosomes and COPII to the ER [7]. In mammalian cells, GTP-Arf1 stimulates phospholipase D, thereby generating phosphatidic acid and thus enhancing local budding [7, 9]. Thus, besides binding Arf1, coatomer interacts with cytosolic tails of receptors and couples the coat assembly with the packaging of cargo molecules [8].

Budding at the ER has recently been reviewed by Springer et al. [10]. At the plasma membrane, budding of vesicles involves five or more independent pathways, as has been reviewed by Riezman et al. [11]. In clathrin-dependent budding, in synaptic vesicle recycling, and perhaps in other pathways such as those involving caveolae, the GTPase dynamin seems to rapidly assemble at the neck areas of budding vesicles to promote budding and/or vesicle detachment. Dynamin contains a plekstrin homology domain that binds to PI 4,5-bisphosphate and PI 3,4,5-trisphosphate [12–14] and thus may represent a functionally significant region of the protein.

Adaptins (AP) and connectors associate with the coat proteins and thereby participate in the fine tuning of the organelle- and cargo-specific assembly. APs are heterotetrameric proteins described in more detail in the next section. Their μ-chains interact with receptors that collect in the budding vesicles and, thus, control the contents of the latter. Some of the proteins involved contain modules that participate in multiple intermolecular interactions. One of the established domains in such "bead on the

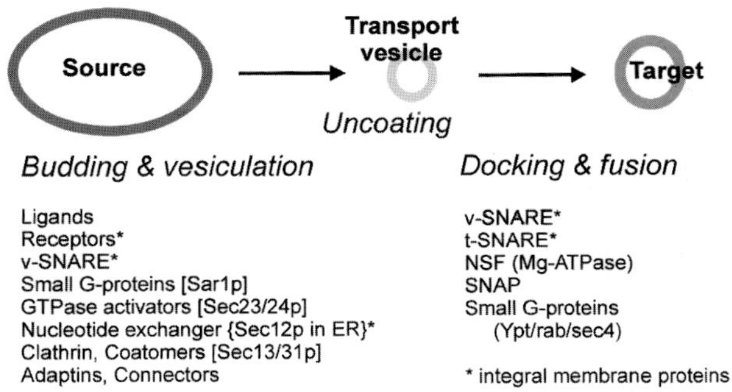

Fig. 2.1. Stages of vesicular transport. In this simplified scheme representative proteins involved in packaging of cargo and targeting of vesicles are listed. Examples referred to in square brackets represent yeast COPII proteins that are necessary and sufficient for formation of vesicles from monolayer liposomes

string" molecules is referred to as epsin homology (EH) and interacts with rab proteins. Epsin binds to an AP (AP-2) via a specialized domain [13, 15]. Furthermore, characteristic proteins co-assemble in the budding membrane and provide an address tag dictating the subsequent interaction with target organelles [16]. These are exemplified by SNAP receptors, so-called v-SNARE proteins that interact with complexes of *N*-ethylmaleimide-sensitive soluble factor, NSF, a Mg^{++} ATPase that was discovered in the laboratory of J.E. Rothman during the initial developments in the field of vesicular traffic, and soluble NSF attachment proteins, SNAPs. Docking at the correct target membrane involves an interaction of the above complexes with t-SNAREs, the corresponding receptor proteins on the receiving organelles. The trafficking of the transport vesicles is regulated by a family of Rab/Ypt/Sec4 proteins that are linked to the SNARE complexes and probably act as molecular timers determining the frequency of membrane fusion events [17].

A scheme on the vesicular transport in yeast indicating the participation of paradigmatic proteins in the processes of budding, fission, targeting and fusion is presented in Fig. 2.1. As mentioned above, in these cells in the ER a cytosolic domain of Sec12p facilitates nucleotide exchange in Sar1p. The activated Sar1p recruits a Sec23p/Sec24p subcomplex to the membrane. There are two (in human cells at least five) orthologues of Sec24p. These proteins are likely to participate in the selection of cargo molecules, since knockout of one of them results in a secretion defect of a subset of proteins [18], although the selection likely involves additional proteins. Sec13/Sec31 components of the COPII coat complex promote the fission of the membrane and, thus, exit from the ER. v-SNARE receptors in these vesicles are represented by Bet1p and Bos1p proteins. The subsequent docking of the vesicles involves a t-SNARE Sed5p, a homologue of human syntaxin-5, which interacts with Sec24p [19].

A series of detailed reviews on membranes and sorting has been edited recently by Zerial [20] and an update discussion on coated vesicles is available [21]. In the next section the vesicular pathways delivering to lysosome molecules that are acquired by endocytosis or synthesized in the rough ER are reviewed succinctly. Subsequently, lysosomal delivery of cytosolic proteins via autophagic vesicles and a transmembrane transport are discussed.

2.2.1
Targeting to Lysosomes from the Extracellular Space

Various complex molecules such as asialoglycoproteins, LDL and mannose 6-phosphate-containing ligands are collected on the surface of cells and conveyed to lysosomes via endocytosis. Commonly, the receptors are recycled from endosomes and reutilized, unless they undergo downregulation. An example of *adsorptive endocytosis* involving cation-independent mannose 6-phosphate receptors (MPR) is considered briefly. Phagocytosis involves different receptors and membrane structures and is not discussed here.

Adsorptive endocytosis is initiated by binding of ligands and grouping of cargo-bearing receptors in coated pits. On the cytosolic side of the pits clathrin coats are assembled along with AP-2 adaptors that consist of α_a/α_c, β_2, μ_2 and σ_2 chains [21]. In AP-2, β_2-chains interact with clathrin and μ_2-chains with MPRs. In the cytosolic tails of cation-independent and -dependent MPRs are the amino acid sequences YSKV and YRGV. These represent a motif, namely tyrosine followed at the third subsequent position by a bulky hydrophobic residue, which mediates interactions with the binding sites of μ_2-chains in AP-2 [22–24]. MPRs and other receptors usually traffic through several compartments. To achieve a characteristic distribution and dynamics of the transport, their cytosolic tails are endowed with multiple consensus motifs [25], each of which allows for a whole spectrum of interactions with the vesicular transport machinery. For example, acid phosphatase and MPRs contain within their cytosolic tails several related signals (Wendland et al. in reference 20), yet these proteins may utilize alternative and independently regulated routes. In the cation-dependent MPR there are one common and two separate sites for interactions with AP-1 and AP-2 [25]. Phosphorylation within the cytosolic tail of furin locks this processing proteinase in a cycle between the plasma membrane and the endosomes, whereas the dephosphorylated form is targeted preferentially to the *trans*-Golgi network (TGN) [26]. Primate lentiviruses produce Nef proteins that downregulate CD4 and thus compromise host defense. Nef proteins are targeted into the endosomal/lysosomal pathway in a stepwise fashion. First, they bind both CD4 and APs in clathrin-coated pits to induce endocytosis and, second, using a diacidic motif they connect CD4 to β-COP to route the cargo to lysosomes [27].

The functions of AP-1 and AP-3 are discussed in the next sections. Recently, another AP, AP-4, has been identified in human cells in association with TGN [28]. In different stages of the vesicular transport specific rab proteins are involved and thus provide another level of regulation. To mention but a few, rab5 participates in the targeting to early endosomes that can be stimulated by the oncogene Ha-ras via protein kinase B/akt [29]. Early endosomes frequently fuse with each other involving rab5 on both membranes. The subsequent delivery to late endosomes depends on rab7 [30]. In the endosomal trafficking rab proteins interact with several partners and appear to facilitate the formation of complexes between SNAREs (for reviews see references 31–33). In yeast cells endosomal traffic involves a rab protein, Vps21p, a GTPase that interacts with a complex of a PI 3-phosphate binding protein Vac1p and the endosomal t-SNARE protein Pep12p [34]. In mammalian cells this pathway is impaired by wortmannin, an inhibitor of PI 3-kinase. In yeast this pathway is characterized by *VPS* mutants. As an example, targeting of carboxypeptidase Y is dependent on PI 3-kinase Vps34p and its upstream regulator protein kinase Vps15p [35].

2.2.2 Targeting from TGN to Endosomes

Both cation-dependent and cation-independent MPRs collect in clathrin-coated vesicles that form at the plasma membrane and at the TGN. In the TGN the bulk of newly synthesized phosphorylated lysosomal enzymes are packaged while bound to MPRs. Here, the receptors interact with AP-1 complexes that consist of γ, β_1 μ_1 and s_1 chains [21]. The cytosolic tails of such receptors possess domains, such as tyrosine and dileucine motifs, which interact with numerous APs. The resultant interactions depend on the specific residues that constitute the motif and its environs. The recruitment of AP-1 onto the budding membrane in the TGN depends on the cooperative interaction of the MPR tails and Arf-1 [36]. In cells that exert regulated secretion, the AP-1-dependent sorting of MPRs in clathrin-coated vesicles continues beyond the TGN in immature secretory granules [37]. In order to return to TGN from endosomes, transmembrane proteins, such as the cation-dependent MPR and furin, interact with additional proteins such as PACS-1, that provide a link to clathrin [38], and TIP47, that recognizes an F/Y signal in the tail of the receptor [39].

2.2.3 Targeting from TGN Directly to Lysosomes

Several proteins, e.g. alkaline phosphatase in yeast [40], a lysosomal integral membrane protein LIMP-II, and melanosomal tyrosinase in mammalian cells [41], have been shown to be packaged in the TGN into vesicles that are delivered directly to vacuoles/lysosomes. This route utilizes AP-3 complexes and does not involve endosomal/prevacuolar compartments (Le Borgne and Hoflack edited in reference 20). AP-3 consists of δ, β_3A/β_3B, μ_3A/μ_3B and σ_3A/σ_3B chains, where the designations A and B refer to non-brain and brain tissues, respectively. Whereas AP-3 had been considered to bind novel coat proteins, Dell'Angelica et al. [42] have reported its interaction also with clathrin, in a fashion similar to that of the previously known APs. In a recent study on a salvage retrieval of proteins from the vacuole, a tentative motif for transport from vacuoles to prevacuoles has been identified. When fused to alkaline phosphatase, the motif directs the latter to prevacuoles. In an unusual itinerary, the modified protein travels from the TGN to prevacuoles via vacuoles [43].

2.2.4 Targeting from TGN via the Plasma Membrane

Among the soluble lysosomal enzymes, acid phosphatase illustrates a mannose 6-phosphate-independent targeting mechanism. The acid phosphatase precursor possesses a transmembrane domain and appears to reach the plasma membrane before being shuttled to endosomes via clathrin-coated vesicles. Within endosomes its cytosolic and transmembrane domains are proteolytically removed and the mature enzyme subsequently resides in lysosomes. The YRHV amino acid sequence in the cytosolic tail resembles the tyrosine motifs in MPRs and is necessary and sufficient for targeting to lysosomes [44]. Interestingly, a significant portion of lysosomal enzymes bearing functional recognition markers may be targeted either directly or via endosomes to the plasma membrane, depending on the expression of the cation-dependent MPR [45]. In the extracellular environment the ligands dissociate from this receptor and may be taken up by the same or another cell using the cation-independent MPR. A benign clinical disorder associated with an elevated secretion of mannose 6-phosphate-competent lysosomal enzymes has recently been described. The underlying defect appears to be enhanced packaging of the cation-dependent MPR into secretory vesicles or its delivery to the plasma membrane. In vitro, the aberrant secretion can be normalized by treating cultured cells with antibodies blocking this receptor [46].

Cation-independent MPR and other endocytic receptors dissociate from their ligands upon reaching the acidic environment of the endosome and subsequently return to the sorting compartment or the plasma membrane. MPRs reaching the lysosomes are likely subjected to rapid degradation. However, late endosomes positive for this receptor seem to fuse transiently with lysosomes [47] and it is not clear whether cation-independent MPR that enters the fused organelles is subjected to rapid degradation or recycling.

2.2.5 Targeting from ER or an Early Golgi Compartment to Endosomes, Lysosomes or Granules

It is known that in various non-mesenchymal tissues lysosomal targeting occurs independently of the MPR. Lysosomal targeting in liver and in white blood cells is little affected in I-cell disease, a disorder associated with the inability to phosphorylate lysosomal enzymes. In normal cells derived from these organs the targeting is rather insensitive to NH_4Cl, a manipulation that significantly interferes with the targeting in mesenchymal cells by preventing the unloading of cargo from MPRs. Likewise, the targeting of various myeloid granule proteins is rather insensitive to NH_4Cl. In HL-60 promyelocytes, immunohistochemistry demonstrates little if any myeloperoxidase in Golgi cisternae, suggesting that its transport from the ER to the azurophilic granule may be aberrant [48]. Nauseef [49] has observed that the proteolytic maturation of myeloperoxidase that occurs after the packaging is rather insensitive to monensin, an ionophore believed to perturb transport through Golgi stacks. Myeloperoxidase becomes phosphorylated, but this modification is not involved in targeting the enzyme into granules. The role of its prosequence in targeting has been examined in two partially complementary studies. Andersson et al. [50] have examined the fate of myeloperoxidase constructs lacking the prosequence when expressed in hematopoietic cells. The mutant enzyme free of the prosequence is no longer targeted to granules but instead is partially degraded and partially secreted. These findings suggest that the prosequence of myeloperoxidase contains a signal required for proper targeting to the azurophilic granule.

Bening et al. [51] expressed in CHO cells a fusion protein consisting of the prosequence of myeloperoxidase joined to the amino terminus of a glycosylated mutant lysozyme construct. The resultant fusion protein was partially targeted to lysosomes in a monensin-insensitive fashion. In addition, the intracellular form possessed exclusively high-mannose oligosaccharides, whereas the secreted form bore complex oligosaccharides.

Owing to the presence of complex type oligosaccharides, as reported for some granular enzymes such as cathepsin G [52], packaging of some granular enzymes in a distal Golgi compartment is rather likely. In this and similar studies oligosaccharides resistant to endo-β-*N*-acetylglucosaminidse H have been considered to represent the complex type.

However, it is important to recognize that trimming of oligosaccharides in lysosomes [51], and probably also in granules, may result in forms resistant to digestion with this endoglycosidase. Therefore, in light of this caveat, this analytic approach to characterizing mannose-rich oligosaccharide structures is valid only if performed under conditions favoring their stability, ideally at the time of the delivery into the lytic compartment.

2.2.6
Targeting to Lysosomes from Cytosol by Micro- and Macroautophagocytosis

During starvation cytosolic proteins and whole organelles in several tissues are degraded following sequestration in autophagocytic vacuoles. A similar phenomenon can be induced in yeast by nutritional shifts. Recently, Klionsky and colleagues have described a genetic dissection of the molecules involved in this "Cvt" pathway in yeast, that is distinct from the "Sec"/"Vps" system. According to their model some cytosolic proteins (e.g. aminopeptidase I) may interact via a propeptide [53] with a "Cvt" complex. The latter directs formation of microvesicles that eventually fuse with lysosomes [54]. An important feature of this delivery is that the targeted proteins are taken up in a folded state. One of the proteins participating in the targeting is a myristoylated and palmitoylated vacuolar inheritance factor, Vac8p [55]. A mechanism similar to this is used in lysosomal targeting of larger portions of cytosol, even including mitochondria, when a major metabolic retooling is needed [56]. In like fashion, fructose 1,6-bisphosphatase, an enzyme subjected to a covalent modification upon a nutritional shift from acetate to glucose as an energy source, is delivered from the cytosol to vacuoles.

As suggested by cytochemical observations, another possibility for delivering portions of cytosol into lysosomes within vesicles is by invagination of the lysosomal membrane. The molecular basis of such transport has not been elucidated.

2.2.7
Targeting Coupled to a Trans-Membrane Transport

Most of the intracellular membranes appear to possess systems for trans-membrane transport operating on unfolded proteins or coupled to their unfolding [56]. Dice and coworkers have described a lysosomal targeting of cytosolic proteins containing motifs with a KFERQ consensus, a sequence present within the N-terminal S-peptide of ribonuclease A that is, in resemblance to the complete enzyme, subjected to a nutritionally regulated degradation. They reported participation of lysosome-associated membrane protein lamp-2, a heat shock protein of the hsp70 family, and ATP in the degradation of soluble KFERQ proteins. This new pathway appears to regulate the level of the intracellular inhibitor of the nuclear factor κB [57] and of α_2-microglobulin [58]. It remains to be shown whether this degradation pathway is coupled to unfolding of the substrate proteins. The degradation of incorrectly folded glycoproteins involves their transport into the cytosol, deglycosylation, and eventual degradation by proteasomes. The oligosaccharides released are processed to a Man_5 GlcNAc species that seems to be transported into the lysosomes to be degraded [59, 60].

2.3 Targeting into Different Granules

Neutrophils contain lysosomes, azurophilic, specific and gelatinase granules, which harbor different partially overlapping sets of proteins. It seems established that there are several vesicular and non-vesicular itineraries to lysosomal organelles and the mechanisms by which the neutrophil granule proteins are segregated into the appropriate compartment during biosynthesis are not clear. To address this question, Le Cabec et al. [61] transfected HL-60 cells, a human promyelocytic cell line lacking specific granules, with cDNA encoding a specific granule protein. Under these conditions the specific granule protein was directed into the azurophilic granule, suggesting that neutrophil granule proteins do not possess signals which mediate their targeting to the appropriate granule population but rather the particular destination is dictated exclusively by the repertoire of granule protein genes being expressed at that time. According to this scheme, the azurophilic granule proteins do not share a targeting sequence but rather are simultaneously expressed during the promyelocytic stage of differentiation. Although the signal which retrieves granule proteins from the secretory pathway and targets them intracellularly has not been defined, it is likely that the delivery of granular proteins follows several routes, including that mediated by the recently described AP-3 complex. Recent studies on pigment variants in *Drosophila* and mouse provide a new impetus to deciphering alternative targeting routes and signals that are involved in the biogenesis of the vacuolar/lysosomal compartments [21].

2.4 Summary

The lysosomal apparatus comprises a group of dynamic morphologically and functionally varied organelles that communicate with cytosol through their single unit membrane. These organelles participate in intracellular digestion, defense mechanisms and storage. Components of the membrane, enzymes and substrates of the digestion reach this compartment by several routes. In this review established and tentative pathways of lysosomal targeting are classified into seven categories that are discussed along with recent studies on vesicular transport and the biogenesis of lysosome-like granules.

Acknowledgements. Our work was supported by the Deutsche Forschungsgemeinschaft (SFB 286) and the Fonds der Chemischen Industrie. The author appreciates the help of Ms. Mechthild Zissel.

References

1. Scheckman R, Mellman I (1997) Does COPI go both ways? Cell 90: 197–200
2. Matsuoka K, Orci L, Amherdt M, Bednarek SY, Hamamoto S, Schekman R, Yeung T (1998) COPII-coated vesicle formation reconstituted with purified coat proteins and chemically defined liposomes Cell 93: 263–275
3. Franco M, Boretto J, Robineau S, Monier S, Goud B, Chardin P, Chavrier P (1998) ARNO3, a sec7-domain guanine nucleotide exchange factor for ADP ribosylation factor 1, is involved in the control of Golgi structure and function. Proc Natl Acad Sci U S A 95: 9926–9931
4. Mossessova E, Gulbis JM, Goldberg J (1998) Structure of the guanine nucleotide exchange factor sec7 domain of human arno and analysis of the interaction with arf GTPase. Cell 92: 415–423
5. Roth MG (1999) Snapshots of ARF1: implications for mechanisms of activation and inactivation. Cell 97: 149–152
6. Goldberg J (1999) Structural and functional analysis of the ARF1-ARFGAP complex reveals a role for coatomer in GTP hydrolysis. Cell 96: 893–902
7. Spang A, Matsuoka K, Hamamoto S, Schekman R, Orci L (1998) Coatomer, Arf1p, and nucleotide are required to bud coat protein complex I-coated vesicles from large synthetic liposomes. Proc Natl Acad Sci USA 95: 11199–11204
8. Bremser M, Nickel W, Schweikert M, Ravazzola M, Amherdt M, Hughes CA, Söllner TH, Rothman JE, Wieland FT (1999) Coupling of coat assembly and vesicle budding to packaging of putative cargo receptors. Cell 96: 495–506
9. Stamnes M, Schiavo G, Stenbeck G, Söllner TH, Rothman J (1998) ADP-ribosylation factor and phosphatidic acid levels in Golgi membranes during budding of coatomer-coated vesicles. Proc Natl Acad Sci USA 95: 13676–13680
10. Springer S, Spang A, Schekman R (1999) A primer on vesicle budding. Cell 97: 145–148
11. Riezman H, Woodman PG, Meer G van, Marsh M (1997) Molecular mechanisms of endocytosis. Cell 91: 731–738
12. Jost M, Simpson F, Kavran JM, Lemmon MA, Schmid SL (1998) Phosphatidylinositol-4,5-bisphosphate is required for endocytic coated vesicle formation. Curr Biol 8: 1399–1408
13. Kirchhausen T (1998) Dynamic dynamin lives up to its name. Curr Biol 8: R792–R794
14. Schmid S, McNiven MA, De Camilli P (1998) Dynamin and its partners: a progress report. Curr Opin Cell Biol 10: 504–512
15. Mayer BJ (1999) Endocytosis: EH domains lend a hand. Curr Biol 9: R70–R73
16. Springer S, Schekman R (1998) Nucleation of COPII vesicular coat complex by endoplasmic reticulum to Golgi vesicle SNAREs. Science 281: 698–700
17. Valentijn JA, Jamieson JD (1998) On the role of rab GTPases: what can be learned from the developing pancreas. Biochem Biophys Res Commun 243: 331–336
18. Pagano A, Letourneur F, Garcia-Estefania D, Carpentier JL, Orci L, Paccaud JP (1999) Sec24 proteins and sorting at the endoplasmic reticulum. J Biol Chem 274: 7833–7840
19. Peng R, Grabowski R, De Antoni A, Gallwitz D (1999) Specific interaction of the yeast *cis*-Golgi syntaxin Sed5p and the coat protein complex II component Sec24p of endoplasmic reticulum-derived transport vesicles. Proc Natl Acad Sci USA 96: 3751–3756
20. Zerial M (1998) Editorial overview. Curr Opin Cell Biol 10: 475–476
21. Jackson T (1998) Transport vesicles: coats of many colours. Curr Biol 8: R609–R612
22. Chen HJ, Yuan J, Lobel P (1997) Systematic mutational analysis of the cation-independent mannose 6-phosphate/insulin-like growth factor II receptor cytoplasmic domain. J Biol Chem 272: 7003–7012
23. Le Borgne R, Hoflack B (1998) Protein transport from the secretory to the endocytic pathway in mammalian cells. Biochim Biophys Acta 1404: 195–209
24. Ohno H, Kirchhausen T, Bonifacino JS (1995) Recognition of tyrosine-based sorting signals by the medium chains of clathrin-associated protein complexes. J NIH Res 7: 50–52
25. Höning S, Sosa M, Hille-Rehfeld A, Figura K von (1997) The 46-kDa mannose 6-phosphate receptor contains multiple binding sites for clathrin adaptors. J Biol Chem 272: 19884–19890
26. Molloy SS, Thomas L, Kamibayashi C, Mumby MC, Thomas G (1998) Regulation of endosome sorting by a specific PP2 A isoform. J Cell Biol 142: 1399–1411
27. Piguet V, Gu F, Foti M, Demaurex N, Gruenberg J, Carpentier JL, Trono D (1999) Nef-induced CD4 degradation: a diacidic based motif in nef functions as a lysosomal targeting signal through the binding of β-COP in endosomes. Cell 97: 63–73
28. Dell'Angelica EC, Mullins C, Bonifacino JS (1998) AP-4 a novel protein related to clathrin adaptors. J Biol Chem 274: 7278–7285
29. Barbieri MA, Kohn AD, Roth RA, Stahl PD (1998) Protein kinase B/akt and rab5 mediate ras activation of endocytosis. J Biol Chem 273: 19367–19370

30. Press B, Feng Y, Hoflack B, Wandinger-Ness A (1998) Mutant Rab7 causes the accumulation of cathepsin D and cation-independent mannose 6-phosphate receptor in an early endocytic compartment. J Cell Biol 140: 1075–1089
31. Hughson FM (1999) Membrane fusion: structure snared at last. Curr Biol 9: R49–R52
32. Schimmöller F, Simon I, Pfeffer SR (1998) Rab GTPases, directors of vesicle docking. J Biol Chem 273: 22161–22164
33. Clague MJ (1998) Molecular aspects of the endocytic pathway. Biochem J 336: 271–282
34. Peterson MR, Christopher GB, Scott DE (1999) Vac1p coordinates rab and phosphatidylinositol 3-kinase signaling in Vps45p-dependent vesicle docking/fusion at the endosome. Curr Biol 9: 159–162
35. Wurmser AE, Jonathan DG, Scott D (1999) Phosphoinostide 3-kinases and their FYVE domain-containing effectors as regulators of vacuolar/lysosomal membrane trafficking pathways. J Biol Chem 274: 9129–9132
36. Le Borgne R, Griffiths G, Hoflack B (1996) Mannose 6-phosphate receptors and ADP-ribosylation factors cooperate for high affinity interaction of the AP-1 Golgi assembly proteins with membranes. J Biol Chem 271: 2162–2170
37. Klumperman J, Kuliawat R, Griffith JM, Geuze HJ, Arvan P (1998) Mannose 6-phosphate receptors are sorted from immature secretory granules via adaptor protein AP-1, clathrin, and syntaxin 6-positive vesicles. J Cell Biol 141: 359–371
38. Wan L, Molloy SS, Thomas L, Liu G, Xiang Y, Rybak SL, Thomas G (1998) PACS-1 defines a novel gene family of cytosolic sorting proteins required for trans-Golgi network localization. Cell 94: 205–216
39. Diaz E, Pfeffer SR (1998) TIP47: a cargo selection device for mannose 6-phosphate receptor trafficking. Cell 93: 433–443
40. Cowles CR, Odorizzi G, Payne G, Emr SD (1997) The AP-3 adaptor complex is essential for cargo-selective transport to the yeast vacuole. Cell 91: 109–118
41. Höning S, Sandoval IV, Figura K von (1998) A di-leucine-based motif in the cytoplasmic tail of LIMP-II and tyrosinase mediates selective binding of AP-3. EMBO J 17: 1304–1314
42. Dell'Angelica E, Klumpermann J, Stoorvogel W, Bonifacino JS (1998) Association of the AP-3 adaptor complex with clathrin. Science 280: 431–433
43. Bryant NJ, Piper RC, Weisman LS, Stevens TH (1998) Retrograde traffic out of the yeast vacuole to the TGN occurs via the prevacuolar/endosomal compartment. J Cell Biol 142: 651–663
44. Peters C, Braun M, Weber B, Wendland M, Schmidt B, Pohlmann R, Waheed A, Figura K von (1990) Targeting of a lysosomal membrane protein. EMBO J 9: 3497–3506
45. Chao HJ, Waheed A, Pohlmann R, Hille A, Figura K von (1990) Mannose 6-phosphate receptor-dependent secretion of lysosomal enzymes. EMBO J 9: 3507–3513
46. Faulhaber J, Fensom A, Hasilik A (1998) Abnormal lysosomal sorting with an enhanced secretion of cathepsin D precursor molecules bearing monoester phosphate groups. Eur J Cell Biol 77: 134–140
47. Bright NA, Reaves BJ, Mullock BM, Luzio JP (1997) Dense core lysosomes can fuse with late endosomes and are re-formed from the resultant hybrid organelles. J Cell Sci 110: 2027–2040
48. Bainton DF (1988) HL-60 cells have abnormal myeloperoxidase transport and packaging. Exp Hematol 16: 150–158
49. Nauseef WM (1987) Posttranslational processing of a human myeloid lysosomal protein, myeloperoxidase. Blood 70: 1143–1150
50. Andersson E, Hellman L, Gullberg U, Olsson I (1998) The role of the propeptide for processing and sorting of human myeloperoxidase. J Biol Chem 273: 4747–4753
51. Bening U, Castino R, Harth H, Hasilik A (1998) Lysosomal segregation of a mannose-rich glycoprotein imparted by the prosequence of myeloperoxidase. J Cell Biochem 71: 158–168
52. Garwicz D, Lindmark A, Persson AM, Gullberg U (1998) On the role of the proform-conformation for processing and intracellular sorting of human cathepsin G. Blood 92: 1415–1422
53. Oda MN, Scott SV, Hefner-Gravink A, Caffarelli AD, Klionsky DJ (1996) Identification of a cytoplasm to vacuole targeting determinant in aminopeptidase I. J Cell Biol 132: 999–1010
54. Klionsky DJ (1998) Nonclassical protein sorting to the yeast vacuole. J Biol Chem 273: 10807–10810
55. Wang YX, Catlett NL, Weisman LS (1998) Vac8p, a vacuolar protein with armadillo repeats, functions in both vacuole inheritance and protein targeting from the cytoplasm to vacuole. J Biol Chem 140: 1063–1070
56. Cleves AE, Kelly RB (1996) Protein translocation: rehearsing the ABCs. Curr Biol 6: 276–278
57. Cuervo AM, Hu W, Lim B, Dice JF (1998) IkappaB is a substrate for a selective pathway of lysosomal proteolysis. Mol Biol Cell 9: 1995–2010
58. Cuervo AM, Hildebrand H, Bomhard EM, Dice JF (1999) Direct lysosomal uptake of alpha 2-microglobulin contributes to chemically induced nephropathy. Kidney Int 55: 529–545
59. Cacan R, Verbert A (1999) Free and N-linked oligomannosides as markers of the quality control of newly synthesized glycoproteins. Biochem Biophys Res Commun 258: 1–5

60. Saint-Pol A, Bauvy C, Codogno P, Moore EH (1997) Transfer of free polymannose-type oligosaccharides from the cytosol to lysosomes in cultured human hepatocellular carcinoma HEPG2 cells. J Cell Biol 136: 45–59
61. Le Cabec V, Cowland JB, Calafat J, Borregaard N (1996) Targeting of proteins to granule subsets is determined by timing and not by sorting: The specific granule protein NGAL is localized to azurophil granules when expressed in HL-60 cells. Proc Natl Acad Sci U S A 93: 6454–6457

Chapter 3

Sorting and Processing of Neutrophil Granule Proteins

Urban Gullberg, Niklas Bengtsson, Elinor Bülow, Daniel Garwicz,
Anders Lindmark, Inge Olsson

3.1 Introduction

As outlined by Dr. Hasilik (Chap. 2), a great deal is known about the various targeting proteins involved in vesicular transport of proteins between different compartments in the cell. This is especially true in the case of lysosomal targeting mediated by the mannose-6-phosphate receptor system. However, other systems must exist and such alternative mechanisms are likely to be operative in myeloid cells, given the data reviewed by Dr. Hasilik.

Below, we review recent information and current theories on the processing and sorting of neutrophil granule proteins. Particular emphasis is put on neutrophil serine proteases and on myeloperoxidase (MPO), proteins important for antimicrobial activity and stored in azurophil granules.

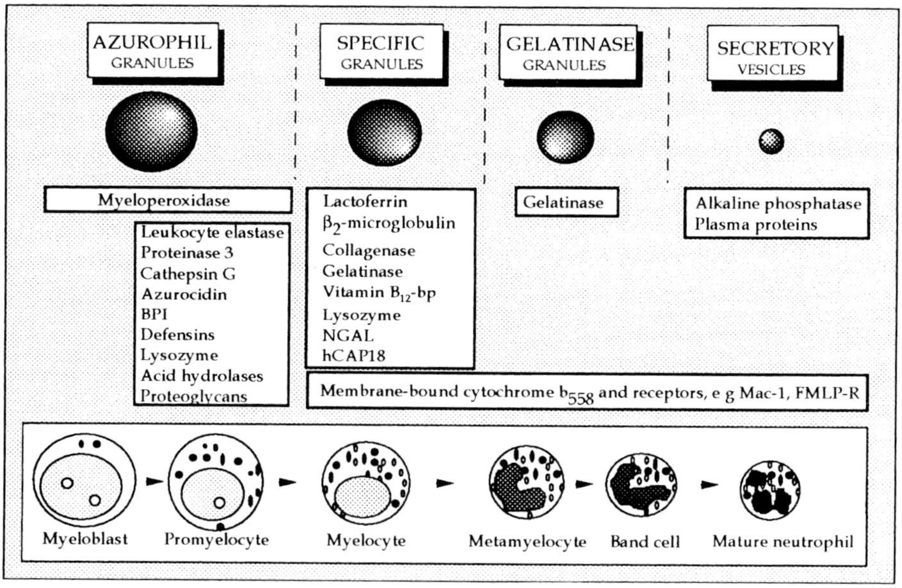

Fig. 3.1. Granule subtypes in neutrophils. The mature neutrophil granulocyte contains four types of granules or granule-like organelles (azurophil, specific, and gelatinase granules, and secretory vesicles) formed during different stages of myeloid differentiation. The granules differ in size, density, and protein content. Modified and reproduced from Gullberg et al. [1] with permission

3.2 Neutrophil Differentiation

The differentiation from progenitor cell to mature neutrophil granulocyte includes morphologically defined stages (Fig. 3.1). The main task of the neutrophil is to migrate to inflammatory foci and there engulf, kill, and degrade invading microbes such as bacteria and fungi. For this purpose, the neutrophil is equipped with a large number of cytoplasmic granules. These are formed in a consecutive manner during maturation of the cell and are traditionally divided into primary (peroxidase-positive) granules, formed mainly during the promyelocyte stage, and secondary (peroxidase-negative) granules, first visible in myelocytes. Primary and secondary granules are also referred to as azurophil and specific granules, respectively, designations which are used in the following text. In addition, relatively recent characterizations have revealed at least two additional storage organelles, gelatinase granules and secretory vesicles. The different granule subtypes show distinct compositions, reflecting their different functional roles (Fig. 3.1) [1, 2].

3.3 Processing and Sorting of Neutrophil Serine Proteases

Cathepsin G, neutrophil elastase, proteinase 3, and the catalytically inactive protease homologue azurocidin, stored in azurophil granules, belong to a superfamily of hematopoietic serine proteases which includes granzymes of cytotoxic T lymphocytes and certain mast cell proteases (Fig. 3.2) [1]. The microbicidal activity of these granule proteins is at least partly independent of their enzymatic activity [3].

Generally, the proteases are synthesized as inactive precursors containing an aminoterminal dipeptide, the removal of which is necessary for enzymatic activation [4]. Longer propeptides may exist for proteinase 3 [5] and azurocidin [6]. Removal of the polar propeptide in pregranule or granule structures allows folding to a mature conformation. Some of the proteases also contain carboxyterminal extensions that are removed during processing. As complex N-linked oligosaccharide side-chains are rather rapidly acquired posttranslationally, newly formed neutrophil serine proteases likely traverse the medial and trans parts of the Golgi, where conversion from high mannose to complex oligosaccharides occurs [1]. The processing and sorting of neutrophil serine proteases are summarized in Fig. 3.3.

Varied amounts of the newly synthesized proforms are secreted from the cell [1]. Interestingly, recent results indicate that secreted proforms may have unique biological effects. The proform of proteinase 3 possesses an ability to decrease the proliferation rate of neutrophil progenitor cells, suggesting a mechanism for negative feedback on myelopoiesis [7].

For investigations of activation and sorting mechanisms, we have utilized the rat basophilic leukemia cell line RBL-1 and the murine myeloblast-like 32D clone 3 cell line in which cDNAs for the proteases can be expressed, leading to normal processing, condensation, and storage of the proteins. Deletion mutants of cathepsin G and neutrophil elastase lacking the carboxyterminal extensions are targeted to granules and converted into catalytically active forms with kinetics similar to those of wild-type enzymes. Therefore, the carboxyterminal prodomains do not seem to be required for

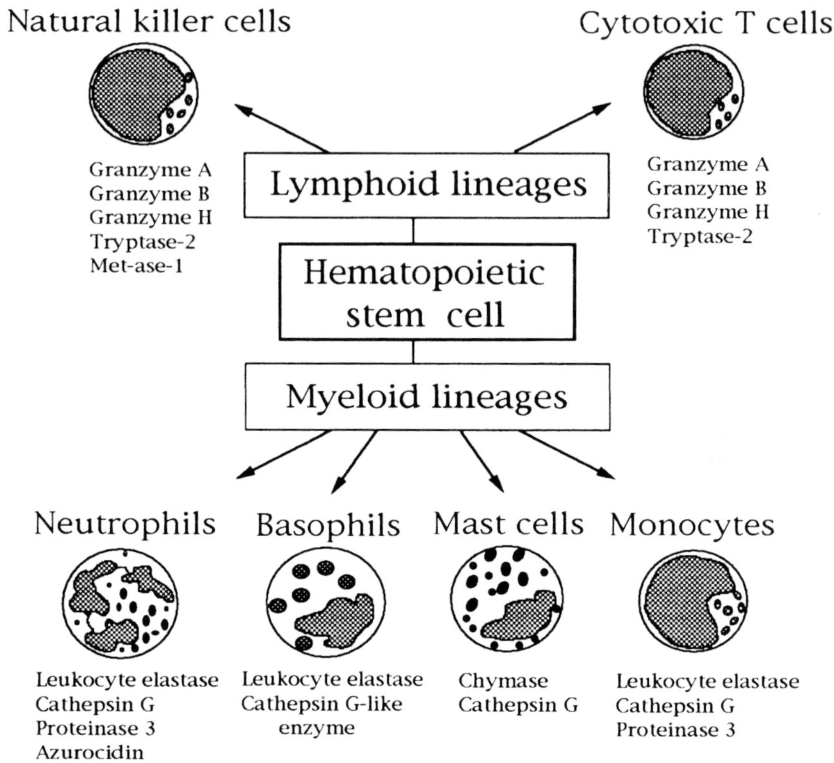

Fig. 3.2. Serine proteases in granules of hematopoietic cells. Cells belonging to lymphoid lineages have a characteristic set of members of the serine protease superfamily stored in their cytoplasmic granules. Cells of myeloid lineages have other members of the serine protease superfamily stored in their granules

targeting or enzymatic activation [8]. Likewise, elimination of the functional glycosylation site of cathepsin G has no obvious consequences for sorting or enzymatic activation [9].

To elucidate the importance of the aminoterminal dipeptide of procathepsin G, its corresponding nucleotides were deleted from cDNA encoding the native protein. It was, however, not possible to stably express the resulting protein either in RBL or in 32D cells [10], suggesting that the dipeptide prevents premature enzymatic activation of the zymogen. Accordingly, an additional mutation was introduced, inactivating the catalytic site, and the resultant mutant was readily expressed as a stable transfectant in RBL and 32D cells and was found to cotranslationally adopt a mature conformation. The conversion into mature conformation was not dependent on removal of the carboxyterminal extension. Although these results question the functional importance of the carboxyterminal processing, they do indicate an important role for the aminoterminal propeptide in keeping the zymogen inactive during intracellular transfer and prior to its activation in the storage compartment [10].

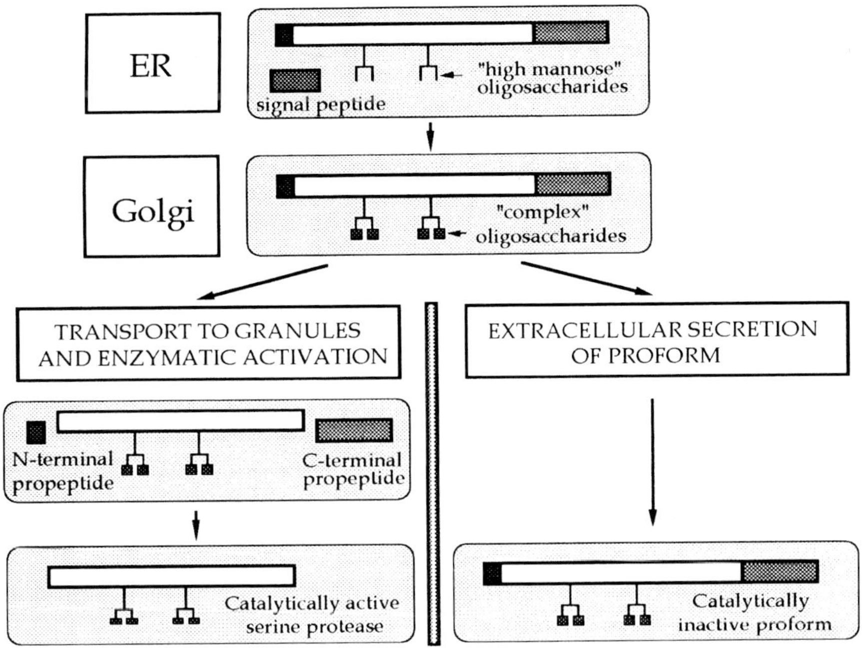

Fig. 3.3. Processing and sorting of the neutrophil serine proteases. After cotranslational cleavage of the signal peptide and asparagine-linked glycosylation in the ER, the oligosaccharide side-chain(s) of the proform is/are converted to complex oligosaccharides in the Golgi apparatus. The protein is then transported to a pregranule/granule compartment, where it is subjected to proteolytic processing (with removal of an aminoterminal propeptide and cleavage of the carboxyterminus), carbohydrate trimming, and storage in a catalytically active form. Alternatively, the protein can be secreted from the trans-Golgi network to the extracellular space in a catalytically inactive proform

3.4 Processing and Sorting of MPO

MPO, stored in azurophil granules, is a very early marker for commitment to myeloid differentiation and is highly expressed during the promyelocyte stage, after which its synthesis wanes. The processing into mature MPO has been characterized (Fig. 3.4) [1, 11, 12]. Newly synthesized MPO precursors are cotranslationally glycosylated to generate heme-free apoproMPO. The subsequent heme insertion is necessary for enzymatic activity and may involve molecular chaperones, such as calreticulin and calnexin [13].

Mature MPO is a homodimer connected by a disulfide bond, each dimer containing a light and a heavy subunit (Fig. 3.4). Following heme insertion, proMPO undergoes extensive proteolytic processing at three sites. A 125 amino acid aminoterminal propeptide, a small peptide between the light and heavy chains, as well as a single serine residue of the carboxyterminus are each excised during processing of proMPO into mature MPO. Similar to the maturation of serine proteases, proteolytic processing occurs in post-endoplasmic reticulum (post-ER) compartments, but mature MPO has been found to be sensitive to digestion with endoglycosidase H (Endo H), indicating

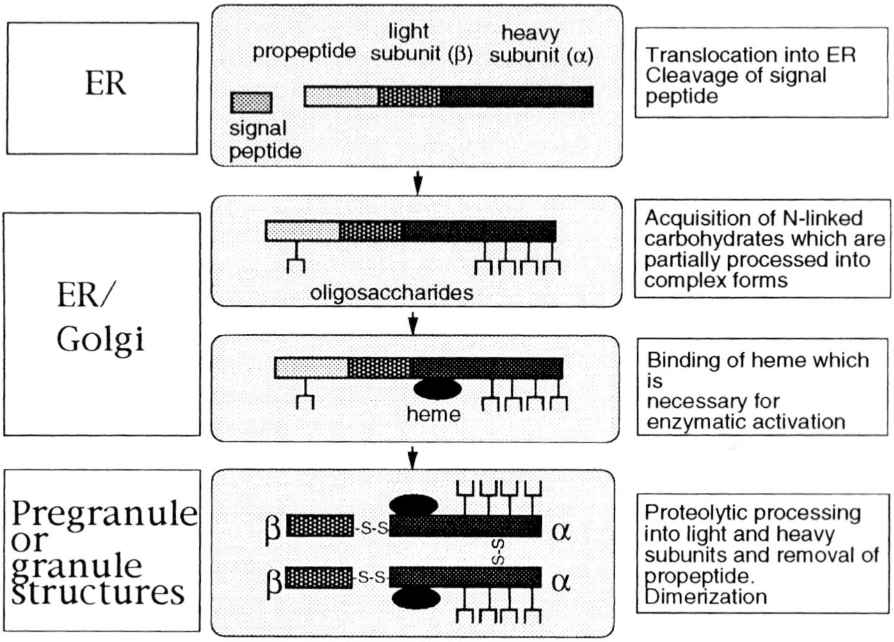

Fig. 3.4. Schematic model for processing of MPO. The signal peptide of the primary translation product is cotranslationally cleaved, followed by asparagine-linked glycosylation to generate apoproMPO. Insertion of heme yields enzymatically active proMPO. During subsequent processing, the aminoterminal propeptide, a small peptide between the light and heavy chains, and a single serine residue at the carboxyterminus are removed. Mature MPO has a dimeric structure

that conversion of oligosaccharides to complex forms does not occur. These and other observations have led to the suggestion that proMPO exits the Golgi at a very early stage, and thus is not transferred through the medial and trans Golgi. However, this interpretation of the data has been questioned (see below, and reference 14).

To explore the function of the aminoterminal propeptide of MPO, we transfected 32D cells with a deletion mutant of preproMPO, lacking the propeptide (MPOΔpro) [14]. Expression of wild-type MPO resulted in normal processing, including insertion of heme, and targeting of MPO to dense cytoplasmic organelles. Although with lower efficiency, MPOΔpro also incorporated heme, indicating that the mutant had capacity for initial folding. However, deletion of the propeptide interfered with further processing; transport to granules and appearance of mature forms were not observed. Rather, a major part of the mutated form was retained in the ER and degraded. The fraction of the protein released from the ER was secreted in an unprocessed form. Thus, the propeptide of MPO may have a role in the targeting of MPO to storage in granules. Retention of the mutant form in the ER is consistent with the observation of a complete sensitivity to Endo H for intracellular forms, while the fraction secreted acquired complete resistance to digestion with the enzyme, indicating transfer through Golgi.

In accordance with earlier results, we have noticed that wild-type mature MPO, stored in granules, retains oligosaccharides sensitive to digestion with Endo H, sug-

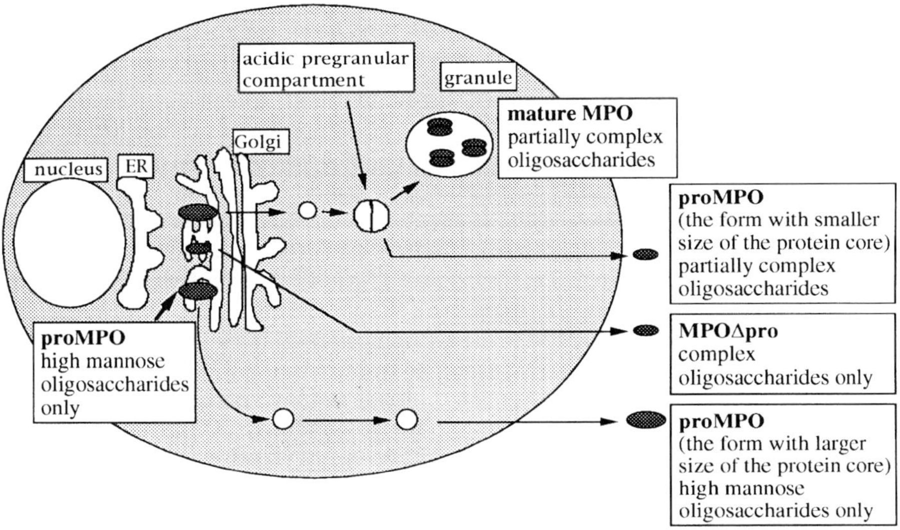

Fig. 3.5. Hypothetical sorting pathways for MPO. Newly synthesized MPO precursor enters and passes Golgi, partly acquiring complex oligosaccharides. In an acidic pregranular compartment processing, including cleavage of the propeptide, takes place. Some of the MPO lacking the propeptide is not retained but escapes to constitutive secretion. An alternative pathway involves early exit from the ER or early Golgi to constitutive secretion of proMPO with retained propeptide and carrying high mannose oligosaccharides only. The putative pathway for MPOΔpro is indicated with some retention in the ER, lack of targeting for storage in granules, and instead constitutive secretion of a product that has acquired complex oligosaccharide side-chains exclusively

gesting exit from early Golgi (see above). However, carbohydrate cleavage with N-glycosidase F results in a further decrease in the molecular weight, demonstrating the presence of carbohydrates also of complex forms (Endo H-resistant), indicating that transfer through Golgi indeed had taken place and that at least some of the stored MPO travels along the same pathway as other azurophil granule proteins. This observation has been confirmed also for endogenous MPO in HL-60 cells. Secreted wild-type proMPO, in contrast to the propeptide deleted mutant forms, also shows oligosaccharides partially of high mannose type. Digestion of secreted wild-type proMPO with N-glycosidase F reveals two distinct protein forms, hypothetically representing the presence or absence of the propeptide, the larger form entirely sensitive to Endo H [14]. These data are compatible with two alternative pathways for constitutive secretion of wild-type proMPO, one exiting from ER or early Golgi and the other leading through the entire Golgi including the trans-Golgi network (TGN) (Fig. 3.5). Hypothetically, one of these pathways includes removal of the propeptide, leading to processing of carbohydrates into some complex forms, while the other one involves protein with retained propeptide resulting in conservation of high mannose oligosaccharides only (Fig. 3.5).

3.5
Is Sorting Dependent on Processing?

Constitutive secretion of both lysosomal and azurophil granule proteins seems to be limited to unprocessed proforms. Therefore, recognition for sorting to granules could be specific for processed forms and processing would then be a prerequisite for sorting to granules. Present data do not, however, support this notion. Unprocessed proforms of MPO are transferred for storage in granules where further processing occurs [1, 12]. Likewise, neutrophil serine proteases are proteolytically processed during transfer for granule storage, but unprocessed forms can nevertheless be found in dense fractions corresponding to granules [1].

Alternatively, and by analogy with the sorting of pro-opiomelanocortin and prorenin, targeting to azurophil granules could be restricted to unprocessed proteins in proform-conformation. This kind of interaction would be possible only for proforms to which certain chaperones or processing enzymes can bind. However, some experimental results do not support this explanation. For instance, creation of a mutant form of cathepsin G, cotranslationally adopting a mature conformation, does not affect its targeting for granule storage, suggesting that the proform-conformation is not a prerequisite for sorting [10]. However, the propeptide of MPO may be important for segregation to granules [14], suggesting that sorting motifs are present in the proform only. In support of this interpretation is a recent report suggesting that the propeptide of MPO can mediate targeting to a lysosomal compartment [15]. In conclusion, however, the various steps in the processing of granule proteins seem most commonly to be post-sorting events rather than a reflection of specific sorting mechanisms.

3.6
Timing of Granule Protein Expression

The synthesis of classes of granule proteins is separated in time, allowing for separate storage without a need for distinct sorting mechanisms for each organelle. Thus, the composition of a given granule reflects the array of genes transcribed and translated into granule proteins at any time, rather than the precise expression of sorting signals to mediate targeting to a special granule subtype. In support of this model are recent data from studies in which neutrophil gelatinase-associated lipocalin (NGAL), a protein that normally distributes to the specific granules, was expressed in promyelocytic HL-60 cells. Transfected NGAL colocalize in the same granules as endogenous MPO, a protein characteristically residing in azurophil granules [16]. These results suggest that MPO and NGAL, proteins normally located in azurophil and specific granules, respectively, utilize common sorting mechanisms, as they are delivered to the same intracellular compartment when coexpressed at the same time in a myeloid cell line.

However, all data are not compatible with this "sorting by timing" hypothesis. Large azurophil granules with only a weak peroxidase reaction are observed rather early in promyelocytes. Later, a collection of small electron-dense azurophil granules with a strong peroxidase reaction appear [1, 2]. However, the defensins are stored in the early-appearing largest azurophil granules despite the fact that their synthesis is initiated after that of other azurophil granule proteins, undermining the model wherein timing of synthesis is the only mechanism responsible for granule diversity.

Additional evidence for distinct sorting mechanisms comes from investigations of promyelocytic NB 4 cells, showing that all proteins of azurophil granules are not produced concomitantly [17]. Moreover, β-glucuronidase, normally residing in azurophil granules, is expressed simultaneously with lactoferrin, a constituent of specific granules. Thus, in spite of concomitant synthesis, the proteins are segregated into distinct storage compartments [17]. Likewise, it has recently been demonstrated that complement receptor type 1 (CR1), which is a marker for the secretory vesicles in neutrophils, when transfected to basophilic RBL cells, is stored in a compartment resembling the secretory vesicles of neutrophils. Importantly, this compartment is distinct from the classical RBL granules, which store proteins such as the hematopoietic serine proteases [18]. Together, these results indicate that differential timing is not the only mechanism for segregation of proteins into distinct granule compartments.

3.7 Sorting by Retention?

What molecular mechanisms retrieve neutrophil granule proteins and target them for storage in granules? It seems likely that proteins destined for storage in granules must be concentrated as cargo already in transport vesicles for further aggregation leading to the condensation state necessary for granule formation. In analogy with sorting of neuroendocrine prohormones [19], a selective aggregation during transfer through the secretory pathway, probably in the TGN, could result in sorting. If heterogeneous aggregates, consisting of several types of granule proteins, are formed, only a few of the proteins have to bind specifically to the TGN membrane and thereby signal for formation and budding of transport vesicles. A similar model has also been proposed for some endocrine hormones, implying that condensed proteins destined for regulated secretion are trapped in immature granules, while soluble proteins are efficiently removed from this compartment for constitutive secretion [20]. Possible mechanisms by which condensation of the content of the vesicle interferes with further secretion and with plasma membrane fusion, and how the vesicles instead fuse to form granules, are not known. Importantly, this hypothetical model eliminates the need for distinct sorting motifs on each type of granule protein.

3.8 Summary

Neutrophil granulocytes are equipped with a collection of cytoplasmic granules destined for regulated secretion, classified as azurophil granules, specific granules, gelatinase granules, and secretory vesicles. Azurophil granules are heterogeneous lysosome-like organelles that store microbicidal proteins such as MPO, the defensins, bactericidal permeability-increasing protein (BPI), and hematopoietic serine proteases. Azurophil granule proteins are subject to extensive posttranslational processing which is often required for catalytic activity. Processing is not a general prerequisite for sorting of the proteins to granules, but elimination of the propeptide of MPO interferes with granular targeting. The synthesis of proteins destined for distinct granules is separated in time, thus offering a mechanism for granule diversity. Separate sorting mechanisms for targeting into each granule type have not been discovered. A selective aggregation of proteins destined for storage in granules would, by intracellular retention, reduce constitutive secretion and facilitate sorting and thus eliminate a need for distinct sorting motifs on each individual granule protein.

Acknowledgements. This work was supported by the Swedish Medical Research Council (Project no. 11546), the Swedish Cancer Foundation, the Alfred Österlund Foundation, the Crafoord Foundation, the Anna-Greta Crafoord Foundation, the Georg Danielsson Foundation, Funds of Lunds Sjukvårdsdistrikt, the John Persson Foundation and Greta and Johan Kock Foundation.

References

1. Gullberg U, Andersson E, Garwicz D, Lindmark A, Olsson I (1997) Biosynthesis and processing of neutrophil granule proteins – insight into neutrophil granule formation. Eur J Haematol 58: 137–153
2. Borregaard N, Cowland JB (1997) Granules of the human neutrophilic polymorphonuclear leukocyte. Blood 89: 3503–3521
3. Levy O (1996) Antibiotic proteins of polymorphonuclear leukocytes. Eur J Haematol 56: 263–277
4. Salvesen G, Enghild JJ (1990) An unusual specificity in the activation of neutrophil serine proteinase zymogens. Biochemistry 29: 5304–5308
5. Garwicz D, Lindmark A, Hellmark T, Gladh M, Jögi J, Gullberg U (1997) Characterization of the processing and granular targeting of human proteinase 3 after transfection to the rat RBL or the murine 32D leukemic cell lines. J Leukoc Biol 61: 113–123
6. Flodgaard H, Østergaard E, Bayne S, Svendsen A, Thomsen J, Engels M, Wollmer A (1991) Covalent structure of two novel neutrophil leucocyte-derived proteins of porcine and human origin. Neutrophil elastase homologues with strong monocyte and fibroblast chemotactic activities. Eur J Biochem 197: 535–547
7. Sköld S, Rosberg B, Gullberg U, Olofsson T (1999) A secreted proform of neutrophil proteinase 3 regulates the proliferation of granulopoietic progenitor cells. Blood 93: 849–856
8. Gullberg U, Lindmark A, Lindgren G, Persson A-M, Nilsson E, Olsson I (1995) Carboxyl-terminal prodomain-deleted human leukocyte elastase and cathepsin G are efficiently targeted to granules and enzymatically activated in the rat basophilic/mast cell line RBL. J Biol Chem 270: 12912–12918
9. Garwicz D, Lindmark A, Gullberg U (1995) Human cathepsin G lacking functional glycosylation site is proteolytically processed and targeted for storage in granules after transfection to the rat basophilic/mast cell line RBL or the murine myeloid cell line 32D. J Biol Chem 270: 28413–28418
10. Garwicz D, Lindmark A, Persson A-M, Gullberg U (1998) On the role of the proform-conformation for processing and intracellular sorting of human cathepsin G. Blood 92: 1415–1422
11. Hasilik A, Pohlmann R, Olsen RL, Figura K von (1984) Myeloperoxidase is synthesized as larger phosphorylated precursor. EMBO J 3: 2671–2676
12. Nauseef WM, Olsson I, Arnljots K (1988) Biosynthesis and processing of myeloperoxidase – a marker for myeloid cell differentiation. Eur J Hematol 40: 97–110
13. Nauseef WM (1998) Insights into myeloperoxidase biosynthesis from its inherited deficiency. J Mol Med 76: 661–668
14. Andersson E, Hellman L, Gullberg U, Olsson I (1998) The role of the propeptide for processing and sorting of human myeloperoxidase. J Biol Chem 273: 4747–4753
15. Bening U, Castino R, Harth N, Isidoro C, Hasilik A (1998) Lysosomal segregation of a mannose-rich glycoprotein imparted by the prosequence of myeloperoxidase. J Cell Biochem 71: 158–168
16. Le-Cabec V, Cowland JB, Calafat J, Borregaard N (1996) Targeting of proteins to granule subsets is determined by timing and not by sorting: The specific granule protein NGAL is localized to azurophil granules when expressed in HL-60 cells. Proc Natl Acad Sci U S A 93: 6454–6457
17. Grégoire C, Welch H, Astaire-Dequeker C, Maridonneau-Parini I (1998) Expression of azurophil and specific granule proteins during differentiation of NB4 cells in neutrophils. J Cell Physiol 175: 203–210
18. Jost C, Klickstei, Wetzler E, Kumar A, Berger M (1998) Intracellular storage and regulated plasma membrane expression of human complement receptor type 1 in rat basophil leukemia cell transfectants. Blood 92: 300–309
19. Bauerfeind R, Huttner WB (1993) Biogenesis of constitutive secretory vesicles, secretory granules and synaptic vesicles. Curr Opin Cell Biol 5: 628–635
20. Kuliawat R, Arvan P (1992) Protein targeting via the "constitutive-like" secretory pathway in isolated pancreatic islets: passive sorting in the immature granule compartment. J Cell Biol 118: 521–529

Chapter 4

X-Ray Crystallographic Studies of Human Myeloperoxidase

C.A. Davey, T. Fiedler, R.E. Fenna

4.1 Introduction

Myeloperoxidase (MPO) occurs in the azurophil granules of mammalian neutrophils where it catalyses the hydrogen peroxide-dependent peroxidation of chloride ion to hypochlorite [1], a potent antibacterial, antiviral and antifungal agent [2]. It is a member of a family of mammalian peroxidases that also includes lactoperoxidase, eosinophil peroxidase and thyroid peroxidase [3]. It is also more distantly related to the prostaglandin endoperoxide synthases [4]. So far, MPO and prostaglandin H (PGH)synthase are the only two members of this family for which X-ray crystal structures have been determined. While it has not been established that the mammalian peroxidases are homologous with plant and fungal heme-containing peroxidases, their mechanisms of catalysis are likely to be quite similar. In all heme peroxidases a histidine residue provides the proximal ligand to the heme iron and a second conserved distal histidine is essential for catalytic activity. With the exception of PGH synthase, all heme-peroxidases, for which three-dimensional structures have been determined, also contain a conserved arginine in the distal heme cavity. While mammalian, plant and fungal peroxidases all catalyze the peroxidation of iodide, MPO is the only one capable of efficient peroxidation of chloride ion.

4.2 Structure Determination

The X-ray crystal structure of human MPO has now been extended to a resolution of 1.75 Å using diffraction data recorded from crystals flash frozen at −180°C. Crystals of human MPO belong to space group $P2_1$ with cell dimensions for the frozen crystals of a = 111.0 Å, b = 63.4 Å, c = 92.2 Å and β = 97.4°C. X-ray diffraction data to a resolution of 1.75 Å have been recorded from a single crystal using a MAR image plate detector mounted on a Rigaku generator equipped with long focusing mirrors and a crystal cooling device. A total of 230 frames of data, each corresponding to a 1.0° rotation of the crystal, were recorded and processed using DENZO and SCALEPACK software [5]. A total of 128,125 unique reflections were obtained from 525,526 intensity measurements with an R_{merge} of 5.3%. This represents 99.8% of the theoretical total number of reflections to a resolution of 1.75 Å. The starting model for refinement was the structure previously determined using diffraction data collected to 2.3 Å at room temperature [6]. Crystallographic refinement was carried out using the X-PLOR [7] program package in combination with manual model building using the TURBO molecular graphics program. The structure has been refined to a crystallographic R-factor of 20.0% for X-ray data in the 30.0 to 1.75 Å range. The R_{Free} for the same resolution range is 24.2%. RMS deviations from ideality for bond lengths and angles are 0.018 Å and 1.43°, respectively.

4.3 Overall Structure

MPO is a dimer in which the two halves are linked by a disulfide bridge between cysteines 153. Each half consists of a large polypeptide of 466 amino acids and a small polypeptide of 108 amino acids, a covalently bound heme, a bound calcium ion, a bound anion and at least three sites of asparagine-linked glycosylation. The overall protein fold (Fig. 4.1) is the same as that reported earlier from a 3-Å resolution X-ray crystal structure of canine MPO [8]. A central core region, that includes the heme is composed of five α-helices, four from the large polypeptide and one from the small, and is surrounded by five globular domains and the small polypeptide chain. One of the glycosylation sites, Asn 317, occurs at the dimer interface, and the conformations of the six individual sugar residues (two N-acetylglucosamines, a fucose and three mannoses) are clearly discernible. The bound calcium ion has typical distorted pentagonal bipyramidal coordination involving ligands from both the large and the small polypeptides. Ligands to the calcium include carboxyl oxygens from each of Asp 96 and Asp 172, hydroxyl oxygens from each of Thr 168 and Ser 174, together with the peptide carbonyl oxygens from residues Asp 96, Thr 168 and Phe 170. There are no water

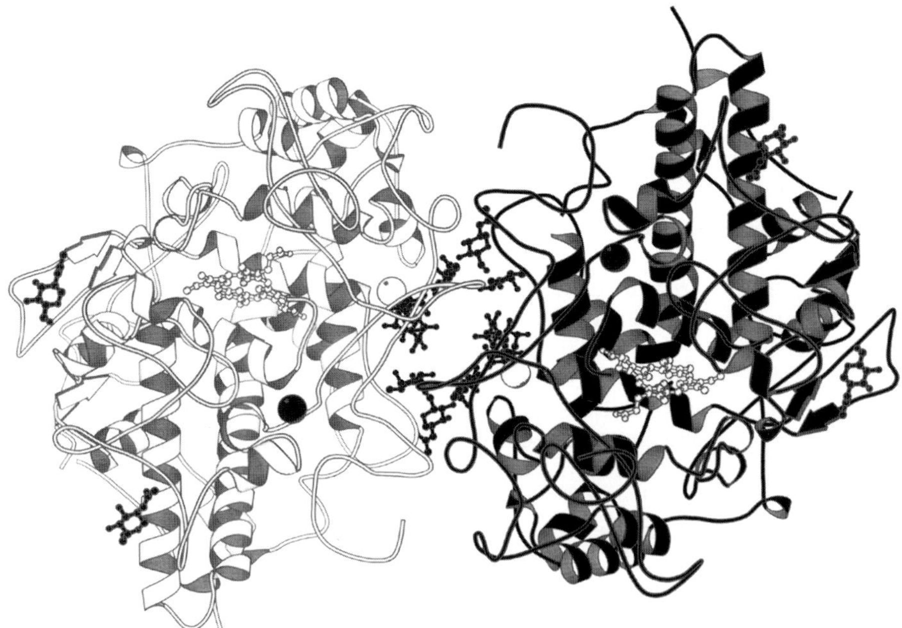

Fig. 4.1. Ribbon diagram of the human MPO dimer. The heme groups are shown in orange, the carbohydrate in blue, the calcium ions in green and bound anions, presumed to be chloride, in yellow. The view is approximately colinear with the twofold axis of symmetry in the molecule

molecules coordinated to the calcium, and the binding site appears to exclude access by solvent molecules. A bound anion has been located near the amino-terminus of the helix containing the proximal His 336. While the binding characteristics of this anion appear consistent with its identification as chloride, additional experiments are required to verify this hypothesis. Since no charged side-chains occur in the binding site, a charge-charge interaction with the helix dipole moment is implicated.

4.4
Heme-Protein Interactions

The structures of the covalent links between the heme and the protein revealed by a 2.3-Å resolution structure of human MPO [6] are confirmed by our more recent studies at higher resolution. The heme in MPO appears to be a derivative of protoporphyrin IX in which the methyl groups on pyrrole rings A and C have been modified in a manner that allows them to form ester linkages with two acidic amino acids (Asp 94 and Glu 242) in the enzyme. In addition, the terminal carbon atom of the vinyl group on pyrrole ring A forms a sulfonium ion linkage with the sulfur atom of Met 243. By analogy to the chemistry of thioether group formation in cytochrome c, it has been proposed that this linkage should involve a bond between the α-carbon of the vinyl group and the methionine sulfur atom [9]. However, as shown in Fig. 4.2, this is

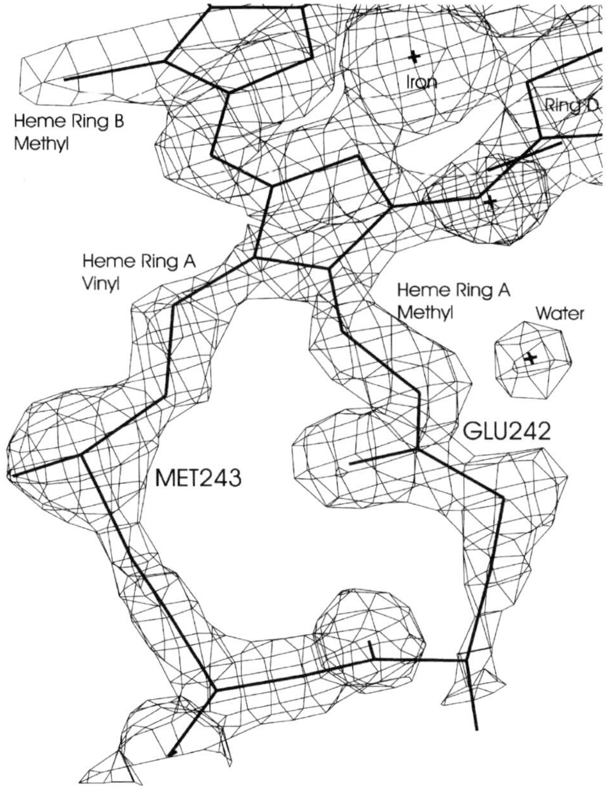

Fig. 4.2. $2F_o$-F_c electron density map at 1.75 Å resolution in the vicinity of the sulfonium ion linkage between the sulfur atom of Met 243 and the terminal carbon of the vinyl group on heme pyrrole ring A

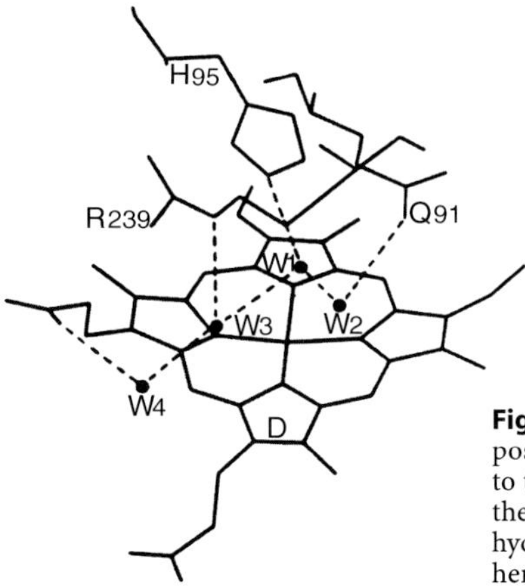

Fig. 4.3. Distal heme cavity of human MPO showing the positions of residues Gln 91, His 95 and Arg 239 with respect to the heme. Four water molecules that occupy this cavity in the native enzyme are also shown. Water molecule W1 is hydrogen bonded to the distal His 95 and is 2.9 Å from the heme iron (from reference 8)

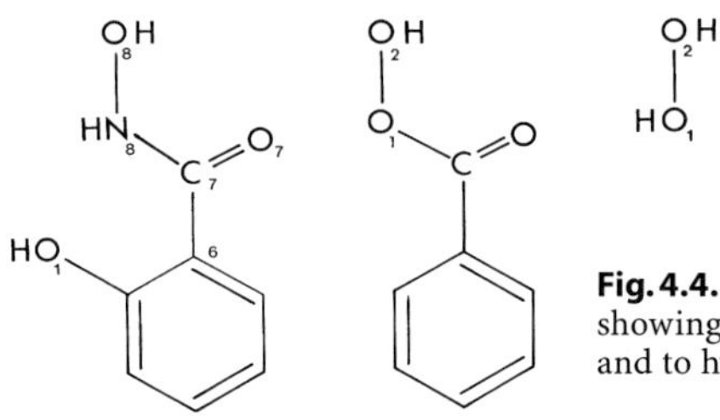

Fig. 4.4. Chemical structure of salicylhydroxamic acid, showing its similarity to the structure of an aromatic peracid and to hydrogen peroxide (from reference 8)

clearly not the case, and it is the terminal carbon atom of the vinyl group that forms a covalent bond with the methionine sulfur atom. The proximal ligand to the heme iron is provided by the NE_2 nitrogen of His 336, while the ND_1 nitrogen of this residue is hydrogen-bonded to the amide oxygen of Asn 421. A cavity on the distal side of the heme is occupied by the side-chains of His 95, Gln 91 and Arg 239 together with four water molecules. As shown in Fig. 4.3, there exists a network of hydrogen bonds between these three residues and the four water molecules. Water W1 is hydrogen-bonded to the NE_2 nitrogen of the distal His 95 and is located roughly mid-way between this nitrogen and the heme iron. A second water molecule (W2) is hydrogen-bonded to the amide group of Gln 91 as well as to W1. Water molecules W3 and W4 are hydrogen-bonded to the guanidinium group of Arg 239 and to a carboxyl group oxygen of the pyrrole ring C proprionate, respectively.

4.5 Salicylhydroxamic Acid Inhibition

Hydroxamic acids are potent inhibitors of mammalian, plant and fungal peroxidases. Salicylhydroxamic acid (SHA; Fig. 4.4) is an analogue of the aromatic peracids that can substitute for hydrogen peroxide in promoting compound I formation in the heme peroxidases [10]. SHA is also representative of a wide range of aromatic alcohols and amines which can act as reducing substrates for compound I [10]. SHA can

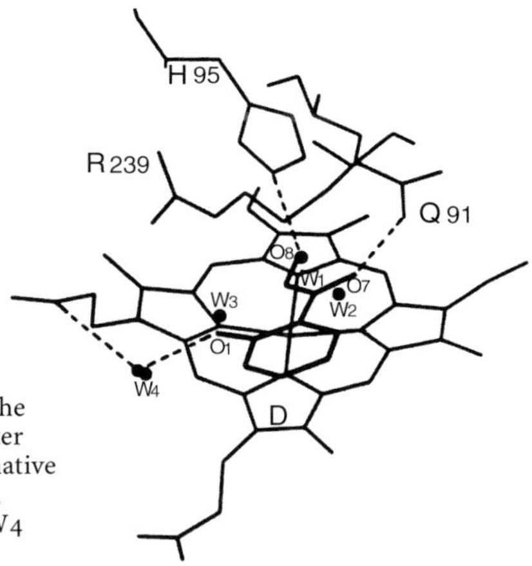

Fig. 4.5. Structure of salicylhydroxamic acid bound in the distal heme cavity of human MPO. The positions of water molecules W1, W2 and W3 that are displaced from the native enzyme by oxygens O_8, O_7 and O_1 of SHA are indicated, together with the small shift (0.4 Å) in the position of W4 (from reference 8)

therefore be regarded as a bi-substrate analogue for MPO. The pH-dependence of SHA binding to MPO ($K_d = 2 \times 10^{-6}\,M$) indicates the involvement of an ionizable group with a pK of about 4. This residue, which has been suggested to be the distal histidine, must be unprotonated for SHA to bind [11, 12].

The X-ray crystal structure of SHA bound to human MPO has been determined to a resolution of 2.3 Å [13]. SHA binds in the distal heme cavity, displacing three water molecules and shifting a fourth by about 0.4 Å from its position in the native enzyme (Fig. 4.5). Apart from the movement of this water molecule, no significant protein conformational changes accompany SHA binding. Interactions between SHA and the enzyme include both hydrogen bonding and Van der Waal's contacts. The hydroxamic acid O_8 hydroxyl group is hydrogen-bonded to the distal His 95 NE_2 (2.6 Å), while the O_7 carbonyl oxygen hydrogen-bonds to the amide group of Gln 91 (2.8 Å). The O_1 ring hydroxyl group of SHA forms a hydrogen bond with the shifted water molecule which is in turn hydrogen-bonded to the heme pyrrole ring C proprionate carboxyl group. The hydroxamic acid N_8 is the closest atom to the heme iron, at a distance of 3.5 Å. The aromatic ring of SHA is inclined at an angle of 20° to the heme, sandwiched between pyrrole ring D and the β, δ and γ carbons of Arg 239. Additional Van der Waal's interactions occur with Phe 99, Phe 366 and Phe 407, each of which has a ring carbon atom within 4.2 to 4.6 Å of the aromatic ring of the SHA molecule.

4.6 Mechanistic Implications

The structure of this complex suggests how an aromatic peracid substrate could interact with MPO to initiate compound I formation. Water molecules W1 and W2 would be displaced from the distal cavity and the hydrogen bonds they make with His 95 and Gln 91 would be replaced by hydrogen bonds to the O_2 hydroxyl and carbonyl group of the peracid. Following deprotonation of the O_2 hydroxyl by the distal histidine, significant movement of this atom must occur for it to interact with the heme iron as an Fe^{III}–OOH transient complex, postulated to occur prior to scission of the O–O bond. A model for this intermediate in cytochrome c peroxidase (CCP) has been proposed on the basis of the structure of an Fe^{II}–O_2 dioxygen complex of the Trp 191 mutant of CCP [14].

The recently determined structure of a complex between horseradish peroxidase (HRP) and benzylhydroxamic acid (BHA) reveals a rather different mode of binding [15]. In this structure, the water molecule located between the distal histidine and the heme iron is not replaced. Instead, the hydroxamic acid group is oriented differently, such that the O_8 hydroxyl group of BHA forms a hydrogen bond with this water molecule rather than eliminating it. This arrangement may be favored because of the ability of the hydroxamic acid N_8H group to form a hydrogen bond with the peptide carbonyl oxygen of Pro 139 and, as noted by the authors, is unlikely to be representative of the mode of interaction for a peracid substrate in which the hydrogen bond donor NH group is replaced by a hydrogen bond acceptor oxygen.

The structure of the MPO-SHA complex also indicates how aromatic alcohols and amines may bind to the enzyme as reducing substrates for compound I. The entrance to the distal cavity is oval in shape, allowing for favorable Van der Waal's interactions with an aromatic ring, oriented approximately parallel with heme pyrrole ring D. In HRP, substrate oxidation has been suggested to occur at the heme edge, close to the δ-meso carbon and the 8-methyl group on pyrrole ring D [16, 17]. In the MPO-SHA complex, the aromatic ring is centered almost directly above the pyrrole ring D 8-methyl group. Single electron reduction of MPO compound I may thus occur by direct electron transfer from the aromatic ring of the reducing substrate to the heme pyrrole ring D rather than by direct interaction with the oxyferryl center of compound I.

Future crystallographic studies on MPO will focus on the identification and characterization of halide binding sites. Since chloride is both a substrate and an inhibitor of the enzyme, there has been considerable speculation as to whether there may be more than one site for halide binding on the enzyme. It is anticipated that such studies will shed light on the mechanism of two-electron oxidation of halide ions by MPO.

4.7 Summary

The X-ray crystal structure of human MPO has been refined at a resolution of 1.75 Å, with an R-factor of 20.0%. The heme group has been characterized as a derivative of protoporphyrin IX in which the methyl groups on pyrrole rings A and C have been modified to allow ester formation with Asp 94 and Glu 242. An additional sulfonium ion linkage occurs between the terminal carbon of the vinyl group on pyrrole ring A and the sulfur atom of Met 243. Structural studies of a complex with the bi-substrate analogue, salicylhydroxamic acid, suggest how both oxidizing peracids and reducing aromatic alcohols and amines could bind to MPO as substrates.

Acknowledgements. This work was supported by a grant (GM49149) from the National Institutes of Health to R.E.F.

References

1. Harrison JE, Schultz J (1976) Studies on the chlorinating activity of myeloperoxidase. J Biol Chem 251: 1371–1374
2. Klebanoff SJ, Clark RA (1978) The neutrophil: function and clinical disorders. North Holland, Amsterdam
3. Kimura S, Ikeda-Saito M (1988) Human myeloperoxidase and thyroid peroxidase, two enzymes both with separate and distinct physiological functions are evolutionarily related members of the same gene family. Proteins Struct Funct Genet 3: 113–120
4. Merlie JP, Fagan D, Mudd J, Needleman P (1988) Isolation and characterization of the complementary DNA for sheep seminal vesicle prostaglandin endoperoxide synthase (cyclooxygenase) J Biol Chem 263: 3550–3553
5. Otwinowski Z (1993) Data collection and processing. In: Proceedings of the CCP4 Study Weekend. Daresbury Laboratory, Warrington, pp 56–62
6. Fenna RE, Zeng J, Davey C (1995) Structure of the green heme in myeloperoxidase. Arch Biochem Biophys 316: 653–656
7. Brunger AT (1992) A system for X-ray crystallography and NMR. X-PLOR Manual (version 3.1). Yale University Press, New Haven
8. Zeng J, Fenna RE (1992) X-ray crystal structure of canine myeloperoxidase at 3 Å resolution. J Mol Biol 226: 185–207
9. Kooter IM, Pierik AJ, Merkx M, Averill BA, Moguilevsky N, Bollen A, Wever R (1997) Difference fourier transform infrared evidence for ester bonds linking the heme group in myeloperoxidase, lactoperoxidase, and eosinophil peroxidase. J Am Chem Soc 119: 11542–11543
10. Schonbaum GR (1973) New complexes of peroxidases with hydroxamic acids, hydrazides and amides. J Biol Chem 248: 502–511
11. Ikeda-Saito M, Shelley DM, Lu L, Booth KS, Caughey WS, Kimura S (1991) Salicylhydroxamic acid inhibits myeloperoxidase activity. J Biol Chem 266: 3611–3616
12. Hori H, Fenna RE, Kimura S, Ikeda-Saito M (1994) Aromatic substrate molecules bind at the distal heme pocket of myeloperoxidase. J Biol Chem 269: 8388–8392
13. Davey CA, Fenna RE (1996) 2.3 Å resolution X-ray crystal structure of the bisubstrate analogue inhibitor salicylhydroxamic acid bound to human myeloperoxidase: a model for a prereaction complex with hydrogen peroxide. Biochemistry 35: 10967–10973
14. Miller MA, Shaw A, Kraut J (1994) 2.2 Å structure of oxy-peroxidase as a model for the transient enzyme:peroxide complex. Struct Biol 1: 524–531
15. Henricksen A, Schuller DJ, Meno K, Welinder KG, Smith AT, Gajhede M (1998) Structural interactions between horseradish peroxidase C and the substrate benzhydroxamic acid determined by X-ray crystallography. Biochemistry 37: 8054–8060
16. Ator MA, Ortiz de Montellano PR (1987) Protein control of prosthetic heme reactivity. Reaction of substrates with the heme edge of horseradish peroxidase. J Biol Chem 262: 1542–1551
17. Ator MA, David SK, Ortiz de Montellano PR (1987) Structure and catalytic mechanism of horseradish peroxidase. Regiospecific meso alkylation of the prosthetic heme group by alkylhydrazines. J Biol Chem 262: 14954–14960

Structural and Biological Properties of Human Recombinant Myeloperoxidase

Nicole Moguilevsky, Alex Bollen

5.1 Introduction

Myeloperoxidase (MPO) is a green hemoprotein present predominantly in azurophilic granules of polymorphonuclear leukocytes (PMN) and blood monocytes. The synthesis of MPO occurs in the promyelocytic stage of myeloid development with production of an 89-kDa precursor that is processed via several intermediates into heavy and light chains of 60 kDa and 15 kDa, respectively. Mature MPO is a homodimer with each half containing a heavy and a light subunit as well as the heme group essential for enzymatic activity.

MPO catalyzes the H_2O_2-dependent oxidation of chloride to hypochlorous acid, a very reactive and microbicidal agent. Thus MPO plays a major role in the antimicrobial activity of PMN and also in damaged tissue at sites of inflammation. The production of recombinant MPO (rMPO) has allowed us to perform functional analyses as well as structural studies in order to better define the nature of the bonds between the heme and the peptide backbone.

5.2 Recombinant MPO

The cDNA encoding MPO has been cloned and characterized [1, 2]. The full-length MPO carrying a signal peptide of 48 amino acids is expressed in dhfr(–)-CHO cells under the control of the RSV (Rous sarcoma virus long terminal repeat) promoter. One recombinant cell line, amplified with methotrexate, overexpresses in the culture medium enzymatically active rMPO up to 14 mg/l per 3 days [3]. Two deletion mutants, coding for truncated signal peptides of 37 and 22 amino acids, respectively, are secreted at the same rate as the full-length molecule. It should be noted that lengths of 48 and 37 residues are rather unusual for eukaryotic signal peptides.

Chromatography of purified product on Q-Sepharose and CM-Sepharose columns resolves two single-chain polypeptides, a major 84-kDa and a minor 94-kDa protein (Fig. 5.1), each existing as a monomer. The N-terminal extremity of the protein species began with Ala 49, thereby indicating that cleavage of the signal peptide occurs between Gly 48 and Ala 49 and that the propeptide is still present.

Analysis of rMPO using glycosidases has indicated that maturation of the carbohydrate side-chains proceeds up to the last stage, i.e. the addition of N-acetylneuraminic acid residues. However, the data shows that rMPO has a higher sugar content than natural MPO (15% versus 12% of the total mass). The excess may reflect an increase in

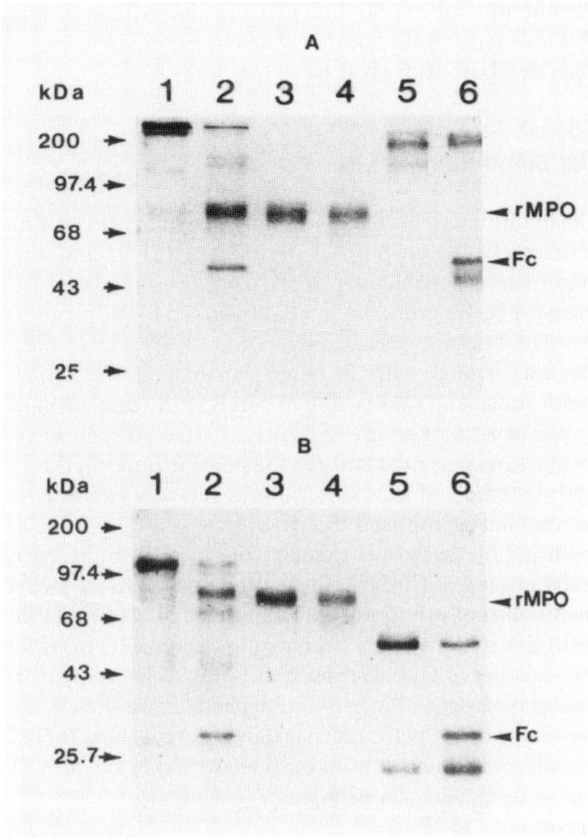

Fig. 5.1A,B. Subunit characterization of rMPO and rMPO-Fc. Purified rMPO and rMPO-Fc were analyzed under non reducing (**A**) or reducing (**B**) conditions, with or without digestion with papain (*lane 1* rMPO-Fc, not digested; *lane 2* rMPO-Fc, digested with papain; *lane 3* rMPO, not digested; *lane 4* rMPO, digested with papain; *lane 5* human IgG, not digested; *lane 6* human IgG, incompletely digested with papain)

mannose content (7.7% versus 4.6%) and suggests that rMPO is not processed as efficiently as the natural enzyme in terms of mannose trimming.

Thus the processing of the rMPO differs from that of wild-type MPO with respect to several features. Recombinant protein expressed in CHO cells undergoes neither proteolytic elimination of the propeptide nor cleavage of the protein into light and heavy chains and fails to dimerize. In order to create a dimer of rMPO, a chimeric antibody-like molecule consisting of the rMPO fused to the second and third constant-sequence (CH2 and CH3) Fc domains of human IgG1 has been constructed (Fig. 5.1) [4]. The fusion protein dimerizes through the Fc fragment while retaining the enzymatic activity of rMPO. In addition, the propeptide of MPO is eliminated during expression of the chimeric construct, suggesting better proteolytic processing of the polypeptide than for rMPO alone.

The specific peroxydasic and chlorinating activities and spectral properties of rMPO, measured with *o*-dianisidine and monochlorodimedon are very similar to those of native MPO. For both enzymes, maximal peroxidasic activity is observed around pH 5.0 in the presence of 200 m*M* NaCl. The chlorinating activity is also dependent on the concentration of hydrogen peroxide. Both enzymes (12 n*M*) at pH

5.5 function optimally when hydrogen peroxide is used at about 10^{-4} to $5\times10^{-5}\,M$ [3, 5]. The thermal stability of rMPO has been shown to be very high, since no loss of activity is detected after 1 h at 60°C [3].

5.3 The Heme-Binding Site

The mammalian peroxidase family of which MPO is a member also includes lactoperoxidase (LPO), eosinophil peroxidase (EPO) and thyroid peroxidase (TPO). The mammalian peroxidases differ from those isolated from plant and fungal sources, since the heme group is covalently linked to the protein matrix. MPO differs from the other mammalian peroxidases, not only by a large red shift of the visible absorption bands associated with the heme prosthetic group, but also because it is the only member of the group that is able to catalyze the oxidation of chloride into hypochlorite.

The crystal structure of dimeric native MPO (resolved at 3 Å) clearly shows several residues in close proximity to the heme group [6], findings confirmed with structural analysis resolved at 2.28 Å [7] and recently extended to 1.75 Å (see Chap. 4). These authors have proposed that the heme is a derivative of protoporphyrin IX wherein modified methyl groups on pyrrole rings A and C form ester linkages with the carboxyl groups of Gln 408 and Asp 260, and where a covalent bond between the vinyl group on ring A and the sulfur atom of Met 409 establishes a sulfonium ion linkage. A similar structure has been proposed for LPO in which the iron protoporphyrin IX is covalently linked via ester bonds between two hydroxylated methyl groups and the corresponding glutamate and aspartate residues, which are conserved throughout the mammalian peroxidase family. A theoretical three-dimensional model for these peroxidases has been built on the scaffold of the MPO X-ray structure [8] and is discussed by Dr. Ferrari (Chap. 15).

Among members of the mammalian peroxidase family, the presence of Met at codon 409 is a feature unique to MPO. The corresponding residues in EPO and TPO are threonine and valine, respectively, whereas glutamine is substituted at this site in bovine LPO. To assess the functional impact of such specificity at residue 409, we have investigated the effect of replacing Met 409 in rMPO by Thr, Val or Gln. These mutations have major effects on the spectroscopic and catalytic properties of the enzyme. The Soret band of the oxidized mutated MPO is found at 412 nm (Table 5.1), which is at a position similar to that found in the corresponding peroxidases. Thus, Met 409 influences significantly the red shift and activity of MPO [9]. However, substituting the Gln in LPO by Met is not sufficient to confer MPO-like properties to LPO [10].

Table 5.1. Optical absorption maximal and specific activity for wild-type and MPO mutants

Protein	Soret peak (nm)	Specific activity (U/mg)
Met409 (MPO)	428	100
Glu409 (LPO)	412	2
Thr409 (EPO)	412	43
Val409 (TPO)	412	3

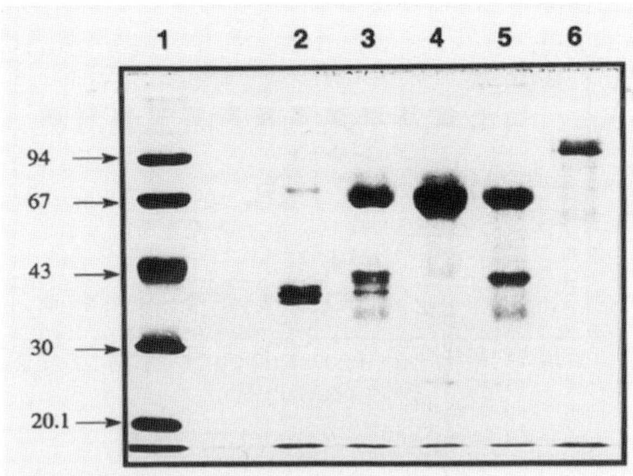

Fig. 5.2. Autocatalytic cleavage of MPO between Met 409 and Pro 410 due to the presence of the sulfonium linkage. Natural, recombinant and mutant MPO (5g) were dissolved in sample buffer without beta-mercapto-ethanol and heated at 90°C for 8 min, and analyzed by SDS-PAGE. *Lane 2* the heating of natural MPO resulted in a fragment of 38 kDa. *Lane 3* under the same conditions the rMPO also showed the presence of a cleavage product of 40–43 kDa, due to the presence of the pro-peptide (108 residues) together with the light and heavy subunit. *Lane 4* the autolytic cleavage was not observed in the mutant Met 409Gln. *Lane 5* as a control, this cleavage occurred in the close mutant Glu 408Gln. *Lane 6* native MPO (not heated)

Taylor et al. [11] have described a unique autolytic cleavage of heated human MPO between Met 409 and Pro 410 of the heavy subunit (55–63 kDa), which generates additional peptide fragments. Under the same heating conditions, the rMPO behaves in a similar fashion (Fig. 5.2) as does the Glu408Gln mutant. However, autolytic cleavage is not observed when Met 409 is replaced by Gln, indicating that autolytic cleavage does not occur. Our studies confirm that Met 409 is involved in a covalent interaction with the chromophore [12, 13].

In addition to the sulfonium ion linkage, the two other linkages of the heme center with Glu 408 and Asp 260 result in considerable distortion of the heme from the planar conformation seen in most heme-containing proteins. Carbonyl stretching vibrations are readily detected by Fourier transform infrared (FTIR) spectroscopy. In proteins carbonyl stretches due to the carboxylic acid chains of Asp or Glu, as well as of esters have been reported. With the help of site-directed mutagenesis and the results of FTIR difference spectroscopy studies, we have provided the first direct spectroscopic evidence for the presence of two distinguishable ester groups for which Asp 260 and Glu 408 are responsible [14, 15].

5.4 Microbicidal Functions of rMPO In Vitro

Purified rMPO has been assayed for cytotoxic effects on various microorganisms. The three essential components of the peroxidase-mediated antimicrobial system consist of hydrogen peroxide (provided by glucose oxidase and glucose), the peroxidase and an oxidizable halide, such as chloride. In vitro, we have shown that rMPO inhibits the growth of *Escherichia coli* [3], *Candida albicans* [16], *Pseudomonas aeruginosa*, *Mycobacterium smegmatis*, *Saccharomyces cerevisiae* (Tournay, personal communication) and human immunodeficiency virus [17, 18]. In this last case, rMPO significantly reduces intracellular viral replication without affecting cell viability.

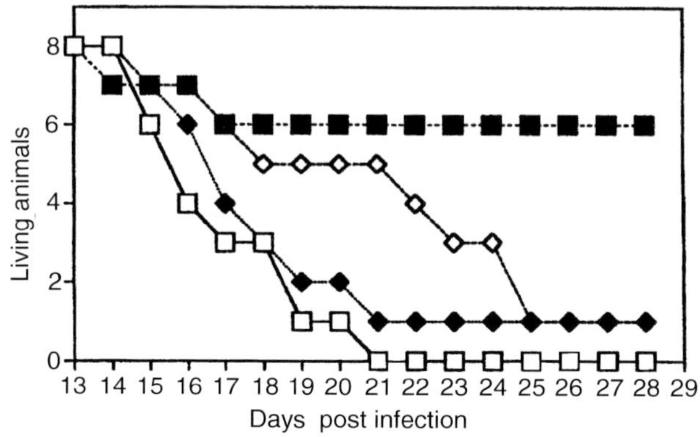

Fig. 5.3. Protection of mice against rickettsiosis by rMPO and rMPO-Fc following infection with 2.3LD50 of *Cowdria ruminantium*. After i.p. injection of a lethal dose of *C. ruminantium*, mice were injected on days 1, 2 and 3 with 8 units/injection of rMPO (◆), 18 units rMPO (■), 8 units rMPO-Fc (◇), or non-injected (□)

5.5 Enhancement of Macrophage Killing Activity

Monocytes, upon maturation into macrophages, gradually lose their MPO with a concomitant decrease in their microbicidal capacity against specific pathogens. However, it has been shown that the interaction of MPO in vitro with macrophages leads to increased macrophage spreading and phagocytosis [19].

We have shown that macrophages can take up rMPO and rMPO-Fc and acquire increased candidicidal activity in vitro [4, 20]. The killing enhancement by rMPO is particularly pronounced with macrophages activated by recombinant human granulocyte-macrophage colony-stimulating factor [21].

Moreover, the fusion of Fc to rMPO further improves the protection conferred by free rMPO in the lethal model of murine rickettsiosis (Fig. 5.3) [4]. Although significant increase in the killing activity of macrophages is conferred by the rMPO-Fc molecule, it does not result from intemalization via the Fc receptor. The binding properties of rMPO-Fc appear to be governed essentially by the highly cationic nature of the protein and seem to be promoted by dimerization.

5.6 Autoantibodies to MPO

Antineutrophil cytoplasmic antibodies (ANCA) are autoantibodies associated with several diseases characterized by a small-vessel vasculitis, as discussed by Drs. Guillevin (Chap. 24), Czernok (Chap. 25), and Tervaert (Chap. 26). MPO is one of the major autoantigens recognized by ANCA. We have compared the antigenicity of native MPO and rMPO using a panel of sera from patients with vasculitis which had previously been shown to contain autoantibodies that bind to native MPO. The aim of the study was to determine whether rMPO might be effective as an extracorporeal immunoabsorbent in vivo, since increasing evidence implicates a pathogenic importance of anti-MPO antibodies.

Using an ELISA, we showed that rMPO is recognized by 56/59 (95%) of the sera that bind to native preparations [22]. The binding was equivalent in 41/59 sera, but 3/59

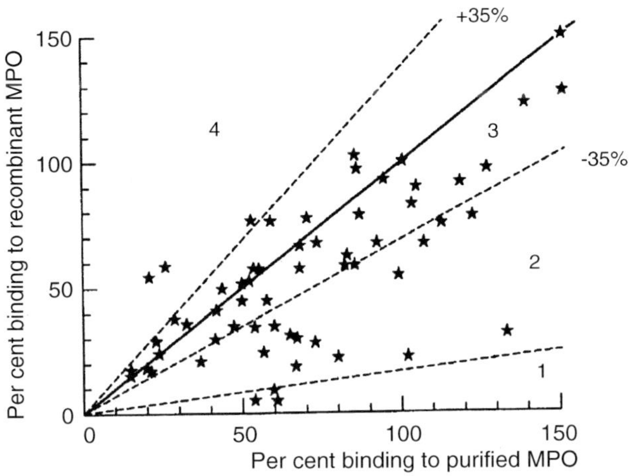

Fig. 5.4. The four binding patterns to recombinant MPO are shown (*group 1* 3 sera that did not recognize rMPO; *group 2* 12 sera with reduced binding to rMPO; *group 3* 41 sera with equivalent binding; *group 4* 3 sera with increased binding to rMPO). Equivalent binding as determined by a mouse monoclonal antibody

sera did not recognize the rMPO and 12/59 had binding reduced by more than 35% with respect to the natural enzyme. Three of the 59 sera showed increased binding to the rMPO (Fig. 5.4).

5.7 Summary

Our recombinant MPO expressed in CHO cells differs from native MPO in several respects. First, it is secreted into the culture medium as an 84-kDa single peptide lacking cleavage between the heavy and light chain and retaining a six amino acid peptide linking the two chains. Second, it does not dimerize, as does the form of MPO recovered from neutrophils. Third, it retains the 14-kDa pro-region, the presence of which may prevent dimerization. Last, although both undergo N-linked glycosylation, rMPO has a slightly higher mannose content.

Despite these differences, rMPO is enzymatically active and possesses spectral properties identical to those of native MPO, thus allowing us to perform structural studies on site-directed mutants of rMPO in which the linkages of the heme to the protein were modified. Increased killing of microorganisms by macrophages loaded with rMPO has also been clearly demonstrated, indicating potential therapeutic use in vitro and in vivo. Finally, rMPO has been shown to be an antigen recognized by 95% of anti-MPO antibodies from patients with systemic vasculitis. Availability of rMPO will be of great value in studies examining the epitopes recognized by anti-MPO antibodies and may provide a new therapeutic approach if used as an extracorporeal immunoabsorbent in vivo.

References

1. Morishita K, Kubota N, Asano S, Kazim Y, Nagata S (1987) Molecular cloning and characterization of cDNA for human myeloperoxidase. J Biol Chem 262: 3844–3851
2. Johnson KR, Nauseef WM, Care A, Wheelock MJ, Shane S, Hudson S, Koeffler HP, Selsted M, Miller C, Rovera G (1987) Characterization of cDNA clones for human myeloperoxidase: predicted amino acid sequence and evidence for multiple mRNA species Nucleic Acids Res 15: 2013–2026
3. Moguilevsky N, Garcia-Quintana L, Jacquet A, Tournay C, Fabry L, Pierard L, Bollen A (1991) Structural and biological properties of human recombinant myeloperoxidase produce by Chinese hamster ovary cell lines. Eur J Biochem 197: 605-614
4. Tournay C, Courtoy PJ, Marodi L, Totte P, Werenne J, Jacquet A, Garcia-Quintana L, Bollen A, Moguilevsky N (1996) Uptake of recombinant myeloperoxidase, free or fused to Fc, by macrophages enhances killing activity toward micro-organisms. DNA Cell Biol 15: 617–624
5. Jacquet A, Deby C, Mathy M, Moguilevsky N, Deby-Dupont G, Thirion A, Goormaghtigh E, Garcia-Quintana L, Bollen A, Pincemail J (1991) Spectral and enzymatic properties of human recombinant myeloperoxidase: comparison with the mature enzyme. Arch Biochem Biophys 291: 132–138
6. Zeng J, Fenna RE (1992) X-ray crystal structure of canine myeloperoxidase at 3 Å resolution. J Mol Biol 226: 185–207
7. Fenna R, Zeng J, Davey C (1995) Structure of the green heme in myeloperoxidase. Arch Biochem Biophys 316: 653–656
8. De Gioia L, Ghibaudi E M, Laurenti E, Salmona M, Ferrari RP (1996) A theoretical three-dimensional model for lactoperoxidase and eosinophil peroxidase, built on the scaffold of the myeloperoxidase X-ray structure. J Biol Inorg Chem 1: 476–485
9. Jacquet A, Garcia-Quintana L, Deleersnyder V, Fenna R, Bollen A, Moguilevsky N (1994) Site-directed mutagenesis of human myeloperoxidase: further identification of residues involved in catalytic activity and heme interaction. Biochem Biophys Res Comun 202: 73–81
10. Watanabe S, Varsalona F, Yoo Y-C, Guillaume J-P, Bollen A, Shimazaki K, Moguilevsky N (1998) Recombinant bovine lactoperoxidase as a tool to study the heme environment in mammalian peroxidases. FEBS Lett 441: 476–479
11. Taylor KL, Pohl J, Kinkade JM Jr (1992) Unique autolytic cleavage of human myeloperoxidase. J Biol Chem 267: 25282–25288
12. Kooter IM, Moguilevsky N, Bollen A, Sijtsema NM, Otto C, Wever R (1997) Site-directed mutagenesis of Met243, a residue of myeloperoxidase involved in binding of the prosthetic group. J Biol Inorg Chem 2: 191–197
13. Kooter I M, Moguilevsky N, Bollen A, Veen LA van der, Otto C, Dekker HL, Wever R (1999) The sulfonium ion linkage in myeloperoxidase: direct spectroscopic detection by isotopic labeling and effect of mutation. J Biol Chem 274: 26794–26802
14. Kooter IM, Pierik AJ, Merkx M, Averill BA, Moguilevsky N, Bollen A, Wever R (1997) Difference Fourier transform infrared evidence for ester bonds linking the heme group in myeloperoxidase, lactoperoxidase, and eosinophil peroxidase. J Am Chem Soc 119: 11542–11543
15. Kooter I M, Moguilevsky N, Bollen A, Sijtsema N M, Otto C, Dekker H L, Wever R (1999) Characterisation of the Asp94 and Glu242 mutants in myeloperoxidase, the residues linking the heme group via ester bonds. Eur J Biochem 264: 211–217
16. Tournay C, Majerus P, Pourtois M, Moguilevsky N, Bollen A (1993) Candidicidal activity of the human recombinant myeloperoxidase system. Antimicrob Agents Chemother 37: 142–143
17. Moguilevsky N, Steens M, Thiriart C, Prieels J-P, Thiry L, Bollen A (1992) Lethal oxidative damage to human immunodeficiency virus by human recombinant myeloperoxidase. FEBS Lett 302: 209–212
18. Chochola J, Yamaguchi Y, Moguilevsky N, Bollen A, Strosberg AD, Stanislawski M (1994) Virucidal effect of myeloperoxidase on human immunodeficiency virus type 1 -infected T cells. Antimicrob Agents Chemother 38: 969–972
19. Lincoln JA, Lefkowitz DL, Cain T, Castro A, Mills KC, Lefkowitz SS, Moguilevsky N, Bollen A (1995) Exogenous myeloperoxidase enhances bacterial phagocytosis and intracellular killing by macrophages. Infect Immun 63: 3042–3047
20. Lefkowitz SS, Gelderman MP, Lefkowitz DL, Moguilevsky N, Bollen A (1996) Phagocytosis and intracellular killing of *Candida albicans* by macrophages exposed to myeloperoxidase. J Infect Dis 173: 1202–1207
21. Marodi L, Tournay C, Kaposzta R, Johnston RB Jr, Moguilevsky N (1998) Augmentation of human macrophage candidacidal capacity by recombinant human myeloperoxidase and granulocyte-macrophage colony-stimulating factor. Infect Immun 66: 2750–2754
22. Short AK, Lockwood CM, Bollen A, Moguilevsky N (1995) Neutrophil and recombinant myeloperoxidase as antigens in ANCA positive systemic vasculitis. Clin Exp Immunol 102: 106–111

Chapter 6

Recent Insights into the Biosynthesis and Processing of Human Myeloperoxidase

William M. Nauseef

6.1 Introduction

My laboratory has had a longstanding interest in the molecular and cell biology of human phagocytes, especially in relation to the oxygen-dependent microbicidal activity of neutrophils. Nearly 20 years ago [1] the identification of several individuals whose neutrophils lacked peroxidase activity sparked our interest in the nature of hereditary myeloperoxidase (MPO) deficiency, specifically in identifying the genotypes responsible and in assessing their impact on MPO biosynthesis.

Chapters 2 and 3 nicely set the stage for the discussion in this chapter. As indicated in Chap. 2, glycoproteins synthesized in the endoplasmic reticulum (ER) undergo post-translational modifications which target them for the secretory pathway or for sorting into specific intracellular organelles. As further discussed in Chap. 2, the mannose-6-phosphate receptor system is the best-characterized system for delivery of lysosomal proteins, although additional mechanisms must also exist. In Chap. 3 the general rules outlined in Chap. 2 are applied to myeloid cells, sharply focusing on the intracellular targeting of proteins destined for the azurophilic granules of neutrophils. This chapter refocuses on events in the biosynthesis of granule proteins but with the emphasis on the issue of "quality control" in the ER during the biosynthesis of MPO. Data from our studies of MPO biosynthesis are used illustrate three purposes subserved by ER molecular chaperones: (1) to prevent nonproductive associations between potentially interactive domains; (2) to retain in the ER malfolded proteins; and (3) to participate in the ER-associated degradative pathway involving the cytosolic proteasome.

6.2 Molecular Chaperones and Normal MPO Biosynthesis

Because of the vectorial nature of glycoprotein synthesis in the ER, nascent proteins experience an "aggregation-sensitive" interval during their biosynthesis [2]. As depicted in Fig. 6.1, domains destined to interact in the native conformation of the mature protein are synthesized sequentially, resulting in the emergence in the ER of the proximal partner before the distal member of a given interactive pair has been synthesized. To minimize the potential for aggregation of these exposed domains, the ER contains a variety of molecular chaperones, including BiP, hsp90, Erp72, calreticulin (CRT), and calnexin (CLN) which interact transiently and reversibly with glycoproteins during their synthesis [3]. In the case of MPO biosynthesis, we have demonstrated that CRT and CLN interact sequentially with the glycosylated forms of MPO precursors in the ER [4, 5].

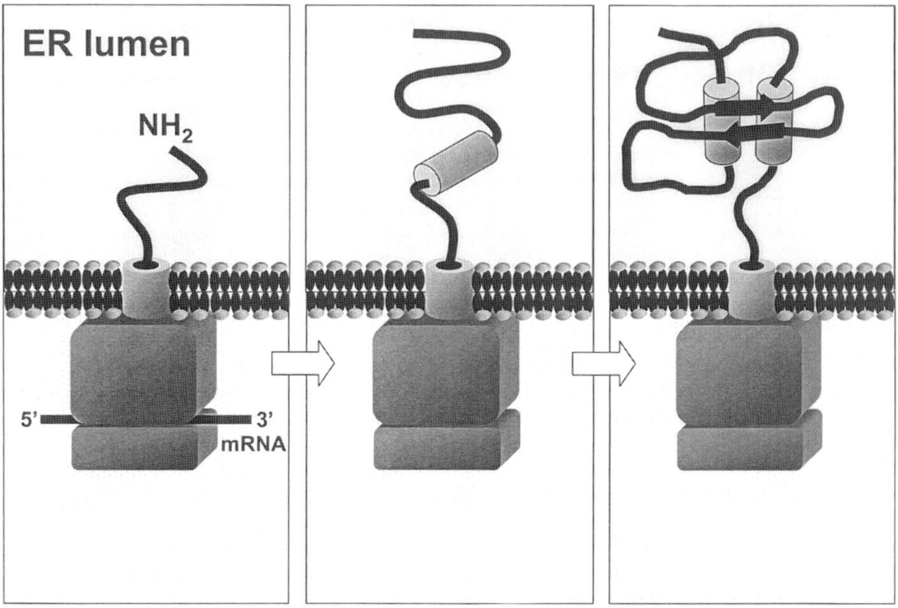

Fig. 6.1. Model of cotranslational folding of a two-domain glycoprotein in the ER. As described by Hartl [2], glycoproteins pass through a period during their biosynthesis when they are especially prone to aggregation. Because of the vectorial nature of protein synthesis, proximal domains in a given protein will be exposed, since their complementary interactive domain has yet to be synthesized

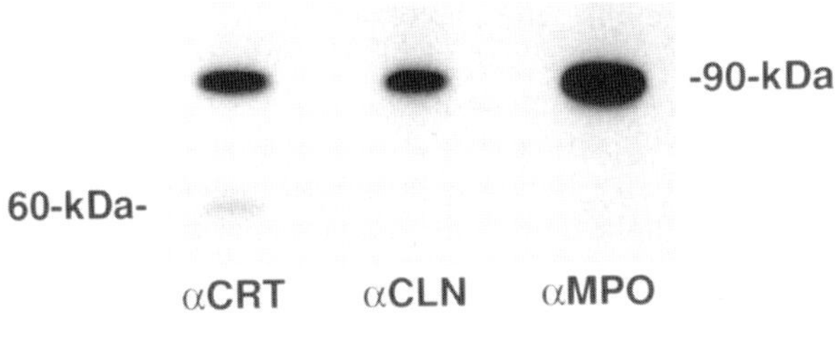

Fig. 6.2. Relative amounts of MPO species associated with calreticulin (CRT) or calnexin (CLN) during biosynthesis by PLB-985 cells. Cells were biosynthetically labelled with ^{35}S-methionine, lysed, and immunoprecipitated under nondenaturing conditions with monospecific antibodies directed against CRT, CLN, or MPO. Immunoprecipitates were washed and subjected to SDS-PAGE and autoradiography. The amounts of radiolabelled myeloperoxidase in each lane were quantitated using a Phosphorimager (Molecular Dynamics). After 2 h of labelling, 27.2% of the total MPO made is associated with CRT, 31.2% with CLN, and the remaining 41.6% free of either molecular chaperone

The population of MPO precursors associated with the molecular chaperone at any given time is assessed by immunoprecipitating biosynthetically radiolabelled MPO from cultured human promyelocytic cells (PLB-985), using antibodies directed against CRT or CLN under nondenaturing conditions. In this way, it is possible to assess what fraction of the MPO precursors made during a specific interval is associated with CRT or CLN. For example (Fig. 6.2), after 2 h of labelling with ^{35}S-methionine, CRT associates with 27% of the MPO precursor synthesized, CLN associates with 31%, and the remainder is free of CRT or CLN. Whereas CRT interacts only with the heme-free apoform of the 90-kDa MPO precursor (apoproMPO) [4], CLN inter-

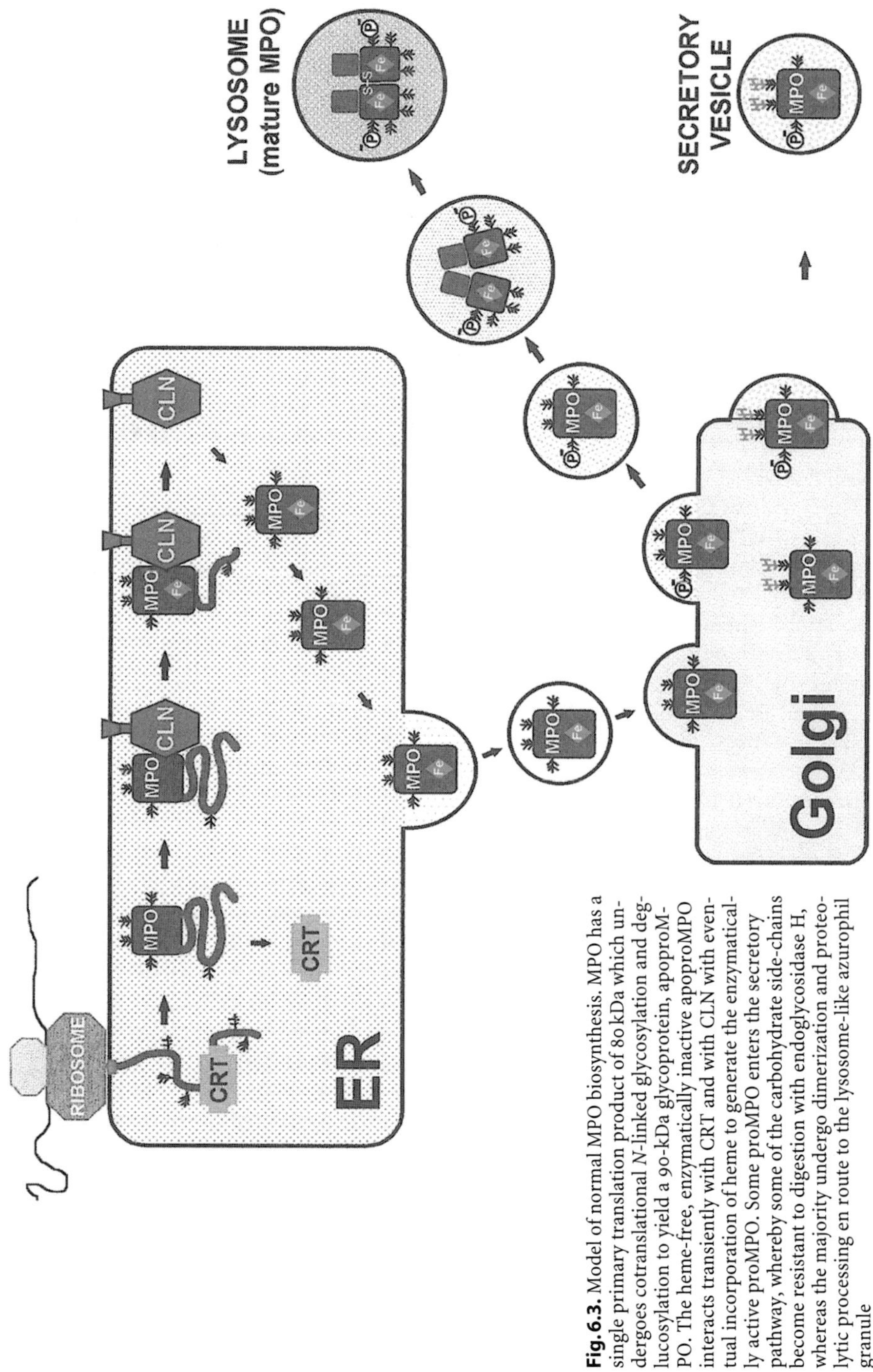

Fig. 6.3. Model of normal MPO biosynthesis. MPO has a single primary translation product of 80 kDa which undergoes cotranslational N-linked glycosylation and deglucosylation to yield a 90-kDa glycoprotein, apoproMPO. The heme-free, enzymatically inactive apoproMPO interacts transiently with CRT and with CLN with eventual incorporation of heme to generate the enzymatically active proMPO. Some proMPO enters the secretory pathway, whereby some of the carbohydrate side-chains become resistant to digestion with endoglycosidase H, whereas the majority undergo dimerization and proteolytic processing en route to the lysosome-like azurophil granule

acts both with apoproMPO and the enzymatically active, heme-containing proMPO [5]. Whether CLN association facilitates insertion of heme into the protein or heme insertion releases CLN from the CLN-apoproMPO complex is not known, but is the subject of ongoing study.

We and others [6, 7, 8] have demonstrated that the interval for maturation of apoproMPO to the mature, lysosomal form of the enzyme is relatively long, in excess of 5 h. Thus apoproMPO and proMPO persist in the ER for a prolonged period of time before proteolytic maturation and organellar targeting occur. Taken together, the results of our studies suggest that CRT and CLN interact with MPO precursors, presumably achieving one of three major functions of chaperones by preventing nonproductive interactions during the period of time apoproMPO and proMPO exist in the ER (Fig. 6.3).

6.3 Genotypes of MPO Deficiency and MPO Structure, Biosynthesis, and Function

To date four genotypes causing missense mutations in MPO have been described. We have examined the impact of several of these mutations on the biosynthesis and processing of MPO using K562 cells transfected with normal or mutant cDNA. This represents a useful system for analysis of the functional impact of specific mutations in MPO, since the cells lack endogenous MPO but are competent to internalize and incorporate heme into hemeproteins [9].

The first mutation identified [10], a replacement of arginine with tryptophan at codon 569 (R569W), has a profound effect on MPO synthesis in the K562 cell transfectants [9]. Although translation and cotranslational glycosylation proceed normally in cells expressing R569W, heme is not inserted into the protein, resulting in failure to generate proMPO or mature MPO subunits. Thus the cells demonstrate a "maturational arrest" at the apoproMPO stage, recapitulating the phenotype seen in patients expressing the R569 W genotype of MPO deficiency.

More recently we have described a missense mutation in which a tyrosine at codon 173 is replaced by a cysteine [11]. Based on the crystal structure of mature MPO [12], discussed in Chap. 4, the light subunit of MPO has an intramolecular disulfide bond between cysteines at position 167 and 180. Thus the Y173C mutation creates the potential for a new intramolecular disulfide bond and concomitant loss of the normal linkage. We predicted that Y173C would likely adversely affect the folding of nascent MPO and thereby interrupt normal processing through the ER. Cells expressing Y173C possess much less peroxidase activity than do cells transfected with cDNA for normal MPO, although the level is clearly above that of the wild-type K562 cells. Using radiolabelled heme precursors, we confirmed that heme is inserted into pro MPO(Y173C), although the radiolabelled species quickly decay after labelling, suggesting either rapid secretion or catabolism of the enzymatically active proMPO.

Biosynthesis of mutant MPO in the Y173C transfectants differs in three ways from that of normal cells in pulse-chase experiments. First, in contrast to cultured promyelocytes expressing endogenous MPO or K562 cells transfected with cDNA encoding normal MPO, the Y173C transfectants fail to secrete any MPO-related proteins into the culture medium. Second, the Y173C proMPO does not undergo proteolytic processing into the normal subunits of mature MPO. Taken together, these two observations coupled with the results of the heme-labelling experiments suggest that Y173C apoproMPO acquires heme in the ER but fails to enter the secretory pathway or the proteolytic processing pathway en route to the azurophilic granule. The third aberrant feature of Y173C MPO biosynthesis is the rapid decay of the 90-kDa precursor during pulse-chase experiments. Thus the proMPO synthesized by cells expressing Y173C is neither secreted nor processed to mature enzyme but rather destroyed in the cell.

6.4 ER Quality Control in Cells Expressing Y173C

Molecular chaperones not only prevent potentially unproductive interactions but also retain in the ER proteins misfolded as a result of mutations [13]. To assess the interactions of CRT and CLN in the processing of Y173C, biosynthetically radiolabelled cells expressing Y173C were immunoprecipitated with antibodies to CRT or CLN under nondenaturing conditions. Both the complexes containing CRT-associated MPO precursors and CLN-associated MPO precursors decay at a slower than normal rate in cells expressing Y173C. The effect is more dramatic with CLN-MPO complexes, as demonstrated when lysates of radiolabelled cells were subjected to sequential immunoprecipitations, first under nondenaturing conditions using anti-CLN antiserum followed by denaturation and immunoprecipitation with anti-MPO (Fig. 6.4). Thus it appears that in the case of the Y173C mutation, CLN exhibits the second feature of molecular chaperones in the ER mentioned above, namely the retention of aberrantly folded proteins.

The observed retention of Y173C by CLN in the ER explains the failure of MPO-related proteins to enter the secretory pathway or to be targeted to the azurophilic granule. We reasoned that the rapid disappearance of Y173C protein from the cell lysate without concomitant processing to mature protein or secretion into the culture medium reflects ER-associated degradation (ERAD) [14]. Because much of the ERAD

Fig. 6.4. Comparison of pulse-chase labelling of K562 cells transfected with cDNA encoding normal or Y173C forms of MPO. K562 transfectants expressing normal MPO (wt-MPO) or mutated MPO (MPOY173C) were labelled biosynthetically for 1 h with ^{35}S-methionine, lysed, and immunoprecipitated under nondenaturing conditions with monospecific antibodies directed against MPO or CLN. The immune complexes from the CLN immunoprecipitation were subsequently denatured, the SDS diluted to 0.2%, and a second immunoprecipitation with MPO antiserum was performed to recover CLN-associated MPO species. In the case of transfectants expressing normal MPO, the CLN-MPO complexes dissociate within the first 2 h. In contrast, CLN remains associated with Y173C MPO precursor for a prolonged period of the chase

is mediated by the cytosolic 26 S proteasome, one would expect that inhibition of proteasome activity would block degradation of Y173C and result in increased recovery of the protein during pulse-chase experiments. In the presence of the proteasome inhibitor *N*-acetyl-l-leucinyl-l-norleucinal [15] or the proteasome-specific inhibitor lactacystin [16], significantly more Y173C is recovered from cells [11], consistent with a role for the proteasome in degradation of the mutant form of MPO.

The suggestion that ERAD of Y173C involves the proteasome poses a topological problem for the cell, as it requires a mechanism by which the existence of a misfolded protein in the ER lumen can recruit the proteasome to the cytoplasmic face of the ER. An identical problem exists in the case of the Z variant of α_1-antitrypsin deficiency, for which Perlmutter and colleagues have clearly demonstrated an important role for CLN [17]. Although CRT and CLN display striking structural and functional homologies, they differ in the means by which they are retained in the ER [3]. CRT is a luminal protein, retained in the ER by virtue of a specific receptor for the C-terminal tetrapeptide, lysine-aspartic acid-glutamic acid-leucine (KDEL). In contrast, CLN is a transmembrane protein with 89 amino acids, including three sites for casein kinase-dependent phosphorylation which extend into the cytoplasm [18]. It is possible that the phosphorylation state of CLN may reflect interactions of the luminal domains of CLN and such signals mediate events on the cytoplasmic face of the ER, including the recruitment of the proteasome.

6.5 Summary

It is noteworthy that the fates of R569W and Y173C are strikingly different, although both result in similar phenotypes, i.e. peroxidase-deficient neutrophils with immunochemical evidence of residual precursor protein. In contrast to studies of the Y173C mutation, we have no data implicating the proteasome in handling of the malfolded R569W protein. How the quality control system distinguishes between normal and malfolded proteins and responds differently to specific errors in folding are unanswered questions currently under study.

Like cystic fibrosis and the Z form of α_1-antitrypsin deficiency, the Y173C variant of hereditary MPO deficiency is an example of a disease whose phenotype reflects processes occurring in the ER. Such diseases give clinical importance to understanding the fundamental events in folding of nascent proteins in the ER and the retention there of abnormally folded species. In the case of Y173C, we do not know if ubiquitination is necessary for proteasome-mediated degradation, nor do we know what are the rate-limiting steps in the pathway from identification of aberrant protein to its degradation. Presumably characterization of additional mutations, based not only on those identified in patients with MPO deficiency but also on sites selected based on the crystal structure of the mature protein, will provide additional insights into the structural determinants important in the biosynthesis, processing, and targeting of MPO and other heme-containing proteins.

References

1. Nauseef WM, Root RK, Malech HL (1983) Biochemical and immunologic analysis of hereditary myeloperoxidase deficiency. J Clin Invest 71: 1297–1307
2. Hartl FU (1996) Molecular chaperones in cellular protein folding. Nature 381: 571–580
3. Helenius A, Trombetta ES, Hebert DS, Simons JF (1997) Calnexin, calreticulin and the folding of glycoproteins. Trends Cell Biol 7: 193–200
4. Nauseef WM, McCormick SF, Clark RA (1995) Calreticulin functions as a molecular chaperone in the biosynthesis of myeloperoxidase. J Biol Chem 270: 4741 –4747
5. Nauseef WM, McCormick SJ, Goedken M (1998) Coordinated participation of calreticulin and calnexin in the biosynthesis of myeloperoxidase. J Biol Chem 273: 7107–71 11
6. Nauseef WM (1986) Myeloperoxidase biosynthesis by a human promyelocytic leukemia cell line: insight into myeloperoxidase deficiency. Blood 67: 865–872
7. Olsson I, Persson A-M, Strömberg K (1984) Biosynthesis, transport, and processing of myeloperoxidase in the human leukemia promyelocytic cell line HL-60 and normal marrow cells. Biochem J 223: 911–920
8 Gullberg U, Andersson E Garwicz D, Lindmark A, Olsson I (1997) Biosynthesis, processing and sorting of neutrophil proteins: insight into neutrophil granule development. Eur J Haematol 58: 137–153
9. Lozzio B Lozzio CB (1979) Properties and usefulness of the original K-562 human myelogenous leukemia cell line. Leuk Res 3: 363–370
10. Nauseef WM, Brigham S, Cogley M (1994) Hereditary myeloperoxidase deficiency due to a missense mutation of arginine 569 to tryptophan. J Biol Chem 269: 1212–1216
11. DeLeo FR., Goedken M, McCormick SJ, Nauseef WM (1998) A novel form of hereditary myeloperoxidase deficiency linked to endoplasmic reticulum/proteasome degradation. J Clin Invest 101: 2900–2909
12. Zeng J, Fenna RE (1992) X-ray crystal structure of canine myeloperoxidase at 3 Å resolution. J Mol Biol 226: 185–207
13. Kopito RR (1997) ER quality control: the cytoplasmic connection. Cell 88: 427–430
14. Brodsky JL, McCracken AA (1997) ER-associated and proteasome-mediated protein degradation: how two topologically restricted events came together. Trends Cell Biol 7: 151–156
15. Jensen TJ, Loo MA, Pind S, Williams DB, Goldberg A, Riordan JR (1995) Multiple proteolytic systems, including the proteasome, contribute to CFTR processing. Cell 83: 129–135
16. Fenteany G, Schreiber SL (1998) Lactacystin, proteasome function, and cell fate. J Biol Chem 273: 8545–8548
17. Qu D, Teckman JH, Omura S, Perlmutter DH (1996) Degradation of a mutant secretory protein, alpha$_1$-antitrypsin Z, in the endoplasmic reticulum requires proteasome activity. J Biol Chem 271: 22791–22795
18. Wong HN, Ward MA, Alexander W, Chevet E, Bains S, Blackstock WP, Solari R, Thomas DY, Bergeron JM (1998) Conserved in vivo phosphorylation of calnexin in casein kinase II sites as well as a protein kinase C/proline-directed kinase site. J Biol Chem 273: 17227–17235

Targets for Myeloperoxidase-Dependent Bactericidal Activity

Henry Rosen

7.1 Introduction

Myeloperoxidase (MPO) is an abundant constituent (up to 5% of dry weight) of polymorphonuclear leukocytes [1]. In conjunction with phagocyte-generated hydrogen peroxide and any of several cofactors, the enzyme catalyzes the formation of a broad array of microbicidal products. The number of potentially microbicidal products exceeds the several effective cofactors implicated, including the halides chloride, bromide, iodide, the pseudohalide thiocyanate, and nitrite [1, 2]. The in vivo relevance of the cofactors based on their anticipated physiologic concentrations is open to debate. Chloride, however, is clearly available at the greater than 10 m*M* concentrations required for it to be an effective MPO cofactor.

7.2 The Hypochlorous/Hypochlorite System as Oxidant

The best-characterized oxidant of the chloride-dependent antimicrobial system is the pair hypochlorous acid/hypochlorite [3]. At physiologic pH, HOCl would be expected to be approximately 50% dissociated so that both hypochlorite and undissociated hypochlorous acid are relevant to consider as antimicrobial factors. Hypochlorous acid may react with free amino groups to form chloramines which, depending on their lipophilicity and molecular size, can either divert oxidants from microbicidal pathways or form less-reactive but longer-lived antimicrobial factors [4, 5]. Partial oxidation of amino acids by HOCl can result in the formation of toxic aldehydes, providing yet another group of antimicrobial factors [6, 7].

The diversity of potential oxidants generated by MPO-mediated reactions does not fully encompass the complexity of potential antimicrobial events. MPO-dependent systems are microbicidal to viruses, bacteria, and fungi (both the yeast and hyphal forms) [1]. The diversity of microbial targets argues strongly for a spectrum of microbicidal mechanisms. For example, viruses, free of their host cells, are metabolically quiescent packets of nucleic acid, a few enzymatic proteins, and a coat of structural proteins that protect the nucleic acids in their latent state. The specifics of MPO-mediated viral inactivation are not well described but presumably relate to chemical modification of this compact array of structural protein and nucleic acid. At the other extreme of complexity, yeasts and hyphal fungi concentrate their genetic nucleic acids in nuclei physically far removed from the cell surface. Mitochondria, responsible for respiratory functions, are similarly dispersed through the cytosol at some distance from the cell surface. Yet MPO-mediated events, presumably generated external to the cell, still are able to prevent replication of treated organisms.

Although bacteria are simpler systems than fungi, factors relevant to MPO-mediated killing remain complex. For example, there are the topology, permeability and metabolic functions of the bacterial envelope to consider. Gram-negative organisms have cell envelopes that differ from those of Gram-positive bacteria. Gram-negative bacteria contain an outer membrane with a specific set of permeability characteristics, a thinner cell wall than most Gram-positive bacteria, and an inner membrane that is the locus of many critical metabolic functions for the cell. These include nutrient transport, ATP synthesis, maintenance of electrochemical and pH gradients, respiration, and electron transport. The cell envelope also plays a role in the initiation of chromosomal DNA synthesis. Just external to the cytoplasmic membrane, the periplasmic space is the locus of activity of a number of cell wall remodeling proteins that covalently bind penicillin, and are designated penicillin-binding protein.

The cell envelope is a significant exclusion barrier for MPO and impedes, to some extent, even small diffusible oxidants such as hypochlorous acid. MPO is too large to be expected to penetrate the outer membrane of bacteria although it has electrostatic properties that cause it to concentrate on the surface of the cell [8]. As MPO-generated oxidants diffuse towards the cell, reactive groups near the surface of the cell exert a "right of first opportunity" to react with and possibly inactivate those oxidants. If these oxidized bacterial molecules are repairable and not immediately critical to the survival of the organism, then the oxidants are consumed without having a lasting antimicrobial effect. On the other hand, if the reactive groups are critical to the immediate survival of the bacterium, the oxidants have contributed to lethal damage.

7.3 Candidate Bacterial Target Molecules

Two considerations are relevant for identifying candidate bacterial target molecules: intrinsic reactivity and location within the cell. Albrich et al. contributed substantially to our awareness of these considerations [9]. Using model compounds of cell envelope lipids and peptidiglycans, they found that such molecules were relatively unreactive towards HOCl. On the other hand, iron sulfur centers, other iron-containing compounds, and sulfhydryl compounds were quite reactive. These structures generally concentrate at the inner membrane of the cell. Sulfhydryl groups are dispersed more widely. These susceptible functional groups, when free in the cellular cytosol, are relatively protected from extracellular HOCl. For example, the enzyme aldolase has a sulfhydryl group that is necessary for its enzymatic activity and is highly susceptible to HOCl inactivation. Using incremental amounts of HOCl, it has been found that microbicidal activity towards *Escherichia coli* substantially precedes any discernible inactivation of cytosolic aldolase. From these studies it was concluded that HOCl exerts a lethal oxidant effect at the surface of the bacterial cell, before it is able to reach cytosolic aldolase. Such observations focused attention principally upon the cell envelope when looking for potential targets for MPO-mediated inactivation.

Of all the many functions that might be readily proposed to be essential to organism survival and propagation, the replication of the genetic information carried in nucleic acids would seem to be the most prominent and important. An attack by the MPO system on *E. coli* nucleic acids would seem at first pass an unappealing hypothesis. The main basis for a reluctance to pursue such a hypothesis would be the perception that these nucleic acids are located in the "center" of the cell and, much like cytosolic aldolase, would be immune from MPO oxidations until the cell had been damaged so extensively that it was no longer viable in any case.

The realization that there are periods during the *E. coli* replicative cycle when chromosomal DNA is closely apposed to the cell envelope during and for several minutes after initiation of chromosomal DNA replication [10–12], led to an investigation of the effects of the MPO system on bacterial DNA synthesis. To our initial surprise and delight, we found that the cell-free MPO system inhibits bacterial DNA synthesis in a fashion that correlates almost exactly with the loss of bacterial viability [13]. Such loss of DNA replication might be "trivial" in the sense that any organism that is dead will have lost its ability to synthesize DNA. However, in contrast to the MPO system, comparable microbicidal systems that are both oxidative (xanthine oxidase/acetaldehyde/Fe-EDTA) and nonoxidative (gentamicin) show significant discordance between lethal effect and DNA synthesis. That is, organisms that had been rendered nonviable by the latter two systems are able to synthesize DNA for significant periods of time before they "give up the ghost".

The xanthine oxidase and gentamicin-dependent antimicrobial systems were selected as being somewhat analogous to the MPO-independent oxidative systems and entirely nonoxidative antimicrobial systems of human neutrophils. It was of interest to see whether the same kinds of differences in the inactivation of DNA synthesis would be observed among human neutrophils that lacked a functional MPO-mediated antimicrobial system. *E. coli* is highly susceptible to nonoxidative microbicidal systems contained in neutrophils [14]. Further, in our hands, and under the experimental conditions employed, normal neutrophils and cells from patients with hereditary complete MPO deficiency and X-linked chronic granulomatous disease (CGD) all kill *E. coli* equally well [15]. Normal cells show a substantial and rapid capacity to stop *E. coli* DNA synthesis during the microbicidal incubation. In this case, however, their is a lag in the sense that at each early time-point there is more bacterial killing than there is inhibition of DNA synthesis. We attribute this difference to the concurrent activities of both the MPO and non-MPO-mediated antimicrobial systems active in phagosomes.

In contrast to normal cells, both MPO-deficient and CGD neutrophils show substantial impairment in their ability to stop *E. coli* DNA synthesis. These defects are substantially corrected by modifying the experimental conditions in a manner designed to restore the MPO-mediated antimicrobial systems of the deficient cells. In the case of MPO-deficient cells, bacteria were coated with MPO prior to phagocytosis. In the case of CGD cells, hydrogen peroxide was continuously generated in the extracellular medium by the action of glucose oxidase on glucose. As a small neutral molecule, hydrogen peroxide had been shown to be able to diffuse through the cell membrane, and presumably across the phagosomal membrane, into the phagosomal compartment. These modifications almost completely correct the deficits in MPO-deficient and CGD cells with respect to stopping *E. coli* DNA synthesis. We concluded that the MPO system is not essential for colicidal activity but that the system is active and a component of the redundant armamentarium of neutrophils.

It also appears to be the system that is most responsible for stopping bacterial DNA synthesis.

As previously noted, there is a period in the replicative cell cycle in *E. coli* during which chromosomal DNA is closely apposed to the cell envelope [10]. The apposed DNA sequence is the site of initiation of DNA replication, a specific DNA sequence of at least 250 bp designated oriC. DNA sequences flanking oriC may also be important for efficient DNA replication. Cell envelopes prepared from *E. coli* demonstrate a sequence-specific capacity to bind oriC, especially when it is in its newly replicated hemimethylated form [11]. Cell envelopes prepared from *E. coli* previously treated with the MPO antimicrobial system progressively lose their capacity to bind hemimethylated oriC [13]. The loss in DNA binding is highly correlated with the decline in microbial viability and in the ability of the organism to synthesize new DNA.

These observations suggest that the impact of the MPO system on bacterial DNA replication is not a general effect on DNA but might be oriC-specific. To evaluate this hypothesis, we compared the replication of plasmid DNA driven by a variety of origins of replication with that of chromosomal DNA following various stages of exposure to the MPO antimicrobial system [16]. Most plasmids showed a significant degree of resistance to MPO-mediated inactivation of plasmid replication, relative to that of the chromosome. The exceptions to this observation were plasmids that had an origin of replication identical to that of the chromosome, so-called minichromosomes, and plasmids that had an origin of replication derived from the P1 phage.

The P1 phage has replicative properties remarkably analogous to those of the *E. coli* chromosome. There is essentially one copy per cell and replication is timed through the cell cycle. The P1 origin has a capacity to bind to the *E. coli* cell envelope in a manner that is very similar to that of the chromosomal oriC [17–19].

The replicon-dependent resistance of plasmid DNA synthesis, compared to chromosomal DNA synthesis, provides indirect information about which features of DNA synthesis are not affected early in the course of MPO action. For example, there must be a sufficiently large high-energy nucleotide pool to allow for DNA assembly, else DNA synthesis would be impeded in a more global fashion. Table 7.1 describes some of the features for DNA replication of the various episomes analyzed. Of these episomes the decline in pCM700 DNA replication is essentially colinear with a decline in chromosomal DNA synthesis. This would be the expected result if the common oriC region of DNA were a critical factor in mediating susceptibility to MPO. Plasmid pUC19, on the other hand, is significantly resistant during early time periods to the effects of MPO. pUC19 replication requires participation of DNA polymerase I, the products of the dnaB, dnaC, dnaG, dnaE and dnaT genes, and also RNA polymerase. Accordingly, we assume that these factors in DNA replication are not those most susceptible to the MPO system.

Plasmid pSC101 is another resistant plasmid, which has an absolute requirement for several functions of the dnaA gene product in its replication. Accordingly, dnaA protein is a less-promising candidate as principal target of MPO attack in mediating microbicidal activity. Plasmid pSP102, with an origin derived from the P1 phage, is significantly more susceptible to the MPO system than other, non oriC plasmids. Another feature that P1 has in common with oriC replicons is the capacity to be bound by the membrane-associated protein that is the product of the SeqA gene. This commonality of regulation of DNA synthesis by the SeqA protein product prompted

Table 7.1. Features of replication properties and protein requirements of plasmids analyzed in reference 16. The table is from that study (*dnaA* protein product of the dnaA gene, *dnaBCGE* protein product of the dnaBCGE gene, *dnaT* protein product of the dnaT gene, *pol I* DNA polymerase I, *RNAP* RNA polymerase, *seqA* membrane associated DNA sequestering protein [12, 19, 22]; ± not all functions of the dnaA protein may be required, determined in part by the impact of diverse dnaA mutations on plasmid replication, *blank cells* incomplete information)

Plasmid	Replicon	Copy no.	Test for linearity (*P*-value)[a]	dnaA[b]	seqA	pol I[b]	dna BCGE[b]	dnaT[b]	RNAP[b]
pCM700	oriC (chromosome)	5–10	0.93	+	+	–			+
pSP102	oriR (P1 phage)	8–10	0.29	±	+	–			
pSC101	(R6-5)	5	<0.0001	+		–	+		
pUC19	pMB (colE1-like)	500–700	<0.0001	±		+	+	+	+
PACYC184	p15a (colE1-like)	10–12	0.0001	±		+	+	+	+

[a] Test for linearity of the relationship between MPO-induced plasmid and chromosomal DNA synthesis. Each *P*-value represents the test of the quadratic term in an orthogonal polynomial regression. One outlying observation in the pSP102 data set is excluded. Low *P*-values support rejection of the hypothesis of linearity

[b] Adapted from reference 21, p 639, Table 18-3

an investigation of the relative susceptibility of *E. coli* SeqA mutants to MPO-mediated inactivation.

Preliminary data indicate that an *E. coli* strain with a SeqA gene disrupted by transposon mutagenesis is somewhat more resistant to the microbicidal effects of the MPO+chloride+glucose+glucose oxidase system. Possibly the failure in SeqA mutants in chromosomal attachment during DNA replication renders these organisms more resistant to MPO. Possibly attachment of this region of DNA might render nucleotides in the polymer susceptible directly to oxidants. Hypohalogenation of unsaturated bonds in membrane lipids might generate epoxides that could alkylate DNA and bind it irreversibly to the membrane thus interfering with the normal replication cycle [20]. Membrane-associated binding proteins other than SeqA may be significantly damaged by MPO in a fashion that effects oriC dependent replication initiation.

At present, it appears that multiple mechanisms are operative simultaneously to affect microbial viability as a consequence of the action of MPO-derived oxidations. Effects on DNA replication are prominent but the mechanisms of replication impairment remain to be elucidated more fully.

7.4 Summary

MPO-mediated antimicrobial systems are one class of an overlapping set of antimicrobial systems accessible to neutrophils and monocytes for host defense against microbial invasion. The complexity of the MPO systems resides, in part, in the diversity of oxidizable cofactors that serve as substrates for formation of potent microbicides. These microbicides probably exert their effects in different ways depending on the nature of the microbe – bacterium, virus or fungus – and the prevalence of decoy targets for oxidation, such as beta-carotenes and nonessential or overabundant sulfhydryl compounds. The topology of potential targets within the microbe, such as location of the DNA replication compartment(s), cell wall-synthesizing enzymes, and electron transport proteins, is also a factor. These issues are considered with respect to the MPO-hydrogen peroxide-chloride antibacterial system acting on *E. coli* with an emphasis on MPO-mediated inhibition of the initiation steps in replication of the bacterial chromosome.

References

1. Klebanoff SJ, Clark RA (1978) The Neutrophil. North-Holland Publishing Co., Amsterdam
2. Klebanoff SJ. (1993) Reactive nitrogen intermediates and antimicrobial activity: role of nitrite. Free Radic Biol Med 14: 351–360
3. Harrison JE, Schultz J (1976) Studies on the chlorinating activity of myeloperoxidase. J Biol Chem 251: 1371–1374
4. Thomas EL (1979) Myeloperoxidase, hydrogen peroxide, chloride antimicrobial system: nitrogen-chlorine derivatives of bacterial components in bactericidal action against *Escherichia coli*. Infect Immun 23: 522–531
5. Thomas EL (1979) Myeloperoxidase-hydrogen peroxide-chloride antimicrobial system: effect of exogenous amines on antibacterial action against *Escherichia coli*. Infect Immun 25: 110–116
6. Selvaraj RJ, Paul BB, Strauss RR, Sbarra AJ (1974) Oxidative peptide cleavage and decarboxylation by the myeloperoxidase-hydrogen peroxide-chloride ion antimicrobial system. Infect Immun 9: 255–260
7. Paul BB, Jacobs AA, Strauss RR, Sbarra AJ (1970) Role of the phagocyte in host-parasite interactions. XXIV. Aldehyde generation by the myeloperoxidase-H_2O_2-chloride system: a possible in vivo mechanism of action. Infect Immun 2: 414–418
8. Selvaraj RJ, Zgliczynski JM, Paul BB, Sbarra AJ (1978) Enhanced killing of myeloperoxidase-coated bacteria in the myeloperoxidase-H_2O_2-Cl- system. J Infect Dis 137: 481–485
9. Albrich JM, McCarthy CA, Hurst JK (1981) Biological reactivity of hypochlorous acid: implications for microbicidal mechanisms of leukocyte myeloperoxidase. Proc Natl Acad Sci U S A 78: 210–214
10. Firshein W (1989) Role of the DNA/membrane complex in prokaryotic DNA replication. Annu Rev Microbiol 43: 89–120
11. Ogden GB, Pratt MJ, Schaechter M (1988) The replicative origin of the *E. coli* chromosome binds to cell membranes only when hemimethylated. Cell 54: 127–135
12. Lu M, Campbell JL, Boye E, Kleckner N (1994) SeqA, a negative modulator of replication initiation in E. coli. Cell 77: 413–426
13. Rosen H, Orman J, Rakita RM, Michel BR, VanDevanter DR (1990) Loss of DNA-membrane interactions and cessation of DNA synthesis in myeloperoxidase-treated *Escherichia coli*. Proc Natl Acad Sci U S A 87: 10048–10052
14. Elsbach P, Weiss J (1983) A reevaluation of the roles of the O_2-dependent and O_2-independent microbicidal systems of phagocytes. Rev Infect Dis 5: 843–853
15. Rosen H, Michel BR (1997) Neutrophil-mediated killing of *Escherichia coli*: redundant contribution of myeloperoxidase-mediated systems. Infect Immun 65: 4173–4178
16. Rosen H, Michel BR, VanDevanter DR, Hughes JP (1998) Differential effects of myeloperoxidase-derived oxidants on *Escherichia coli* DNA replication. Infect Immun 66: 2655–2659
17. Abeles A, Brendler T, Austin S (1993) Evidence of two levels of control of P1 oriR and host oriC replication origins by DNA adenine methylation. J Bacteriol 175: 7801–7807
18. Brendler T, Abeles A, Austin S (1991) Critical sequences in the core of the P1 plasmid replication origin. J Bacteriol 173: 3935–3942
19. Brendler T, Abeles A, Austin S (1995) A protein that binds to the P1 origin core and the oriC 13mer region in a methylation-specific fashion is the product of the host seqA gene. EMBO J 14: 4083–4089
20. Winterbourn CC, Berg JJM van den, Roitman E, Kuypers FA (1992) Chlorohydrin formation from unsaturated fatty acids reacted with hypochlorous acid. Arch Biochem Biophys 296: 547–555
21. Kornberg A, Baker TA (1992) DNA replication. W.H. Freeman and Company, New York
22. Slater S, Wold S, Lu M, Boye E, Skarstad K, Kleckner N (1995) E. coli SeqA protein binds oriC in two different methyl-modulated reactions appropriate to its roles in DNA replication initiation and origin sequestration. Cell 82: 927–936

Reactions of Myeloperoxidase and Production of Hypochlorous Acid in Neutrophil Phagosomes

Christine C. Winterbourn, Christine J. van Dalen, Mark B. Hampton,
Anthony J. Kettle

8.1
Introduction

It is generally accepted that the physiological function of myeloperoxidase is to produce hypochlorous acid. However, this green heme protein does not fit the classical definition of an enzyme. Usually an enzyme is specific in the reaction it catalyzes and in its choice of substrates. In contrast, myeloperoxidase has many and varied substrates, and catalyzes at least three different reactions. This indiscriminate catalytic power argues against a singular role for myeloperoxidase. Rather, it suggests that its function will be modulated by the local milieu. Therefore, we need to fully appreciate the complexity of its reaction mechanisms to understand how it acts within neutrophil phagosomes during microbial killing and how it contributes to inflammation. In addition to hydrogen peroxide and chloride, potential physiological substrates for myeloperoxidase include superoxide, thiocyanate, tyrosine, tryptophan, urate, nitric oxide, and nitrite, and there are bound to be others. Their reactions also need to be considered when assessing the function of myeloperoxidase. In this chapter we outline the redox transformations that myeloperoxidase undergoes during turnover, and give examples of how the various alternative substrates influence its activity. We also take a kinetic modelling approach to predict how myeloperoxidase functions in the neutrophil phagosome and test what parameters are critical for efficient production of hypochlorous acid.

8.2
The Reaction Mechanisms of Myeloperoxidase

8.2.1
Oxidation of Halides and Pseudohalides

Myeloperoxidase is a tetramer that consists of two identical dimers. Each heavy subunit has a heme prosthetic group that contains ferric iron (for a review of myeloperoxidase see reference 1). The native or ferric enzyme (MP^{3+}) reacts rapidly with hydrogen peroxide (k_1 2×10^7 $M^{-1}s^{-1}$) in a reversible reaction to form the redox intermediate compound I (Fig. 8.1, reaction 1) [2]. Compound I is a ferryl π-cation radical in which the iron has a formal oxidation state of 5^+. It has a reduction potential of 1.1 V, which enables it to oxidize halides to hypohalous acids (Fig. 8.1, reaction 2). Chloride is accepted as the physiological substrate because it is present in extracellular fluids at far higher concentrations than are iodide, bromide, and the pseudohalide thiocyanate. However, the specificity constant for thiocyanate is 700 times that for chloride

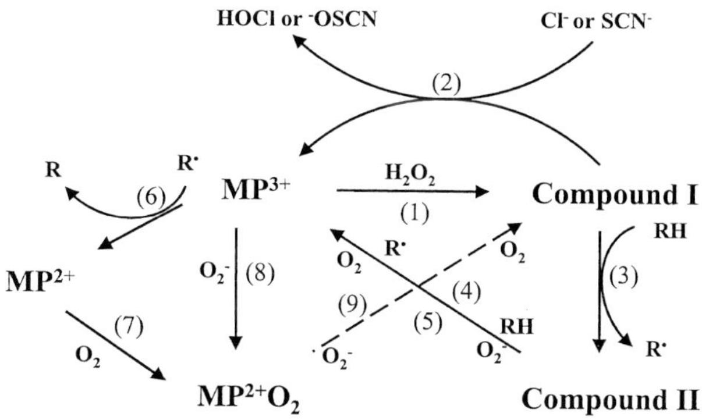

Fig. 8.1. Reactions of myeloperoxidase. RH is an organic substrate such as tyrosine or urate. MP^{3+}, MP^{2+}, and $MP^{2+}O_2$ are native myeloperoxidase, ferrous myeloperoxidase, and compound III, respectively

[3]. At 100 μ*M* thiocyanate and 100 m*M* chloride, about 40% of the hydrogen peroxide used by myeloperoxidase is converted to hypothiocyanite ($^-$OSCN) with the remainder producing hypochlorous acid (HOCl). Therefore, thiocyanate will also be oxidized by myeloperoxidase in vivo. At the millimolar concentrations of thiocyanate that exist in the mouth and the gut, myeloperoxidase should produce only hypothiocyanite. Thus, hypothiocyanite as well as hypochlorous acid should be physiological products of the halogenation cycle of myeloperoxidase (Fig. 8.1, reactions 1 and 2). Little oxidation of bromide and iodide would be expected in extracellular fluid [3].

8.2.2
One-Electron Oxidation of Substrates and Inhibition of Myeloperoxidase

Compound I also oxidizes a multitude of organic substrates (RH), such as tyrosine and urate, to free radical intermediates (R·) with the concomitant formation of compound II (Fig. 8.1, reaction 3). The formal oxidation state of the iron in compound II is 4^+. This redox intermediate is unable to oxidize chloride or thiocyanate but reacts with organic substrates to complete the classical peroxidation cycle (Fig. 8.1, reactions 1, 3 and 4). The rate of peroxidation via this pathway is dependent on the turnover of compound II. Substrates react with it at least ten times more slowly than with com-

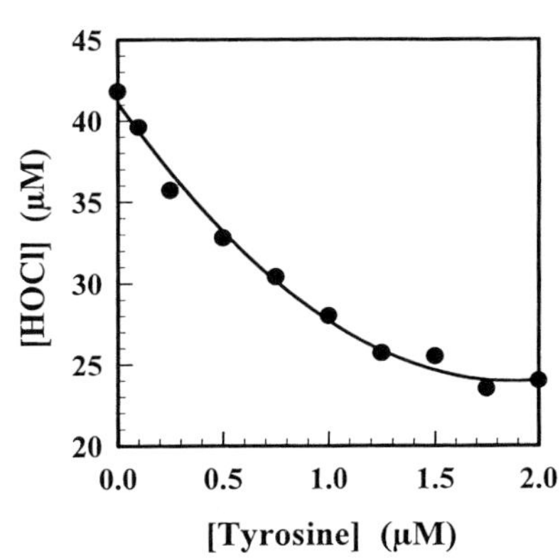

Fig. 8.2. The effect of tyrosine on production of HOCl by myeloperoxidase. Myeloperoxidase (50 n*M*) was incubated in 100 m*M* phosphate buffer (pH 7.4) containing 10 m*M* taurine, 100 m*M* sodium chloride and varying concentrations of tyrosine. Reactions were carried out at room temperature and started by adding 50 μ*M* hydrogen peroxide. After 5 min the concentration of taurine chloramine produced was determined as previously outlined [4]

pound I. The partitioning of hydrogen peroxide between the halogenation cycle and the peroxidation cycle depends on the relative rates at which competing substrates reduce compound I. With 100 m*M* chloride and 100 μ*M* tyrosine, about 95% of the hydrogen peroxide is used to produce hypochlorous acid and 5% to oxidize tyrosine (Fig. 8.2). With 1 m*M* tyrosine about 50% of the hydrogen peroxide is used to oxidize tyrosine.

A number of organic substrates generate radicals that reduce native myeloperoxidase to its ferrous form (MP^{2+}), which rapidly binds oxygen to form oxymyeloperoxidase or compound III (Fig. 8.1, reactions 6 and 7) [5]. For these substrates, the rate of turnover of compound III determines how readily they are oxidized. If the enzyme is trapped at compound III, the substrates are poorly peroxidized.

8.2.3 Reversible Inhibition of Myeloperoxidase

The oxidation of substrates by peroxidases is strongly influenced by their one-electron reduction potentials [6]. Since the one-electron reduction potential of tryptophan is almost identical to that of tyrosine, the two compounds might be expected to react equally well with myeloperoxidase. However, tryptophan is a poor substrate and a good inhibitor of the isolated enzyme (Fig. 8.3). At 2 μ*M*, it inhibits production of hypochlorous acid production by 50%. This occurs because tryptophan reacts rapidly with compound I to convert it to compound II (Fig. 8.1, reaction 3) but reacts slowly with compound II. Thus, tryptophan traps the enzyme in its inactive compound II form. From this result, it is likely that steric factors, as well as reduction potential, determine the extent to which substrates reduce compound II. Tryptophan is unlikely to inhibit myeloperoxidase in vivo because other substrates, such as tyrosine, should cycle compound II (Fig. 8.1, reaction 4).

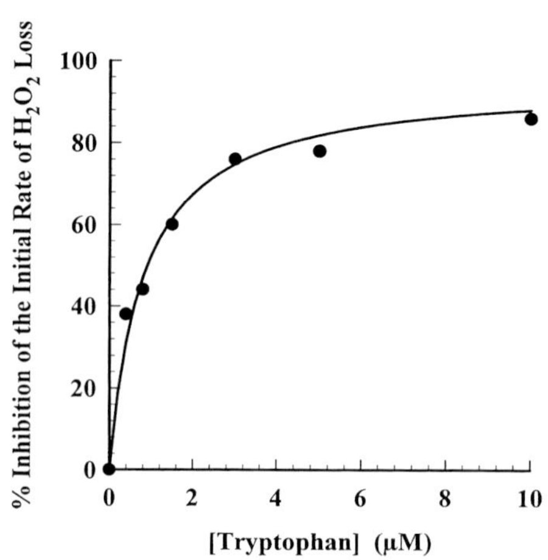

Fig. 8.3. Inhibition of myeloperoxidase by tryptophan. Myeloperoxidase (20 n*M*) was added to 30 μ*M* hydrogen peroxide in 10 m*M* phosphate buffer (pH 7.4) containing 140 m*M* sodium chloride, 300 μ*M* methionine to scavenge HOCl, and varying concentrations of tryptophan. The concentration of hydrogen peroxide was continuously monitored using a hydrogen peroxide electrode [4]. Enzyme activity was measured as the rate of hydrogen peroxide loss over the first 10 s of the reaction. Under these reaction conditions all the hydrogen peroxide consumed by myeloperoxidase is converted to hypochlorous acid [7]

8.2.4
Reactions of Myeloperoxidase with Superoxide

Superoxide also prevents tryptophan and related inhibitors from attenuating production of hypochlorous acid by reducing compound II (Fig. 8.1, reaction 5). It also reacts with the native enzyme to form compound III (Fig. 8.1, reaction 8), which is the dominant redox intermediate of myeloperoxidase in stimulated neutrophils [8]. Compound III must be turned over to compound I for neutrophils to produce hypochlorous acid. The most likely reaction is reduction of compound III by superoxide (Fig. 8.1, reaction 9). This reaction was first proposed by Cuperus et al. [9], and is supported by our kinetic modelling (see below). Compound I is the ultimate product of this reaction. However, a redox intermediate that is electronically equivalent to compound I may be formed and responsible for the hydroxylation activity of myeloperoxidase [10].

8.3
Nitrite as a Peroxidase Substrate for Myeloperoxidase

It has recently been shown that myeloperoxidase oxidizes nitrite and the enzyme may be responsible for much of the nitration of proteins that occurs at inflammatory foci [11, 12]. However, using a hydrogen peroxide electrode to monitor hydrogen peroxide utilization, we have found that at physiological concentrations nitrite is a poor substrate for myeloperoxidase. In fact, it is a very good inhibitor of hypochlorous acid production, with 1 μM nitrite inhibiting by about 50% [13]. Nitrite inhibits production of hypochlorous acid in much the same way as does tryptophan, by rapidly converting compound I to compound II but reacting only slowly with compound II. Thus, myeloperoxidase lingers at compound II and production of hypochlorous acid is retarded. The spectral identification of compound II during the oxidation of nitrite signifies that it must reduce compound I in a one-electron reaction forming nitrogen dioxide in the process (reaction 10). With stimulated neutrophils as the source of active myeloperoxidase, nitrite is not as effective an inhibitor as with the purified enzyme. However, if superoxide dismutase is present, inhibition is much greater because superoxide can no longer recycle compound II (Fig. 8.1, reaction 5).

$$\text{Compound I} + NO_2^- \rightarrow \text{Compound II} + NO_2^{\cdot} \qquad (10)$$

To reconcile our results with the recent findings that myeloperoxidase catalyzes nitration of tyrosyl residues in proteins and peptides, we have investigated the ability of purified myeloperoxidase to nitrate the peptide acetyl-Ser-Gln-Asn-Tyr-Pro-Val-Val-NH_2. We chose this peptide because myeloperoxidase does not oxidize its tyrosyl residue.

Using physiologically relevant nitrite concentrations, we observed very little nitration of the peptide. Only about 1% of the hydrogen peroxide was accounted for by the production of 3-nitrotyrosine in the peptide. However, addition of 10 μM tyrosine increased the efficiency of peptide nitration by nearly 300%. This result indicates that myeloperoxidase requires a cosubstrate such as tyrosine to catalyze nitration of proteins.

Adding tyrosine would enhance nitration because tyrosine radicals generated by myeloperoxidase would exchange with tyrosyl residues on the peptide, which would in turn react at diffusion controlled rates with nitrogen dioxide (reactions 11 and 12).

$$\text{Tyr}^{\cdot} + \text{P-TyrH} \rightarrow \text{Tyr} + \text{P-Tyr}^{\cdot} \qquad (11)$$

$$\text{NO}_2^{\cdot} + \text{P-Tyr}^{\cdot} \rightarrow \text{P-Tyr-NO}_2 \qquad (12)$$

There is also very little nitration of the peptide by stimulated neutrophils. For appreciable nitration to occur, nitrite has to be present at 1 m*M* (Table 8.1). Nitration is dependent on myeloperoxidase because it is virtually abolished by catalase and the heme inhibitor azide. As with the purified enzyme, tyrosine enhances nitration by neutrophil-released myeloperoxidase. Methionine, a scavenger of hypochlorous acid, also increases the formation of 3-nitrotyrosine in the peptide. This latter result indicates that nitration does not occur via nitryl chloride which is formed from the reaction of hypochlorous acid with nitrite [14]. Interestingly, superoxide dismutase increases nitration dramatically. This result rules out peroxynitrite, which is formed in the reaction of superoxide with nitric oxide, as the nitrating agent. Instead, it is likely that superoxide blocks nitration by reacting with tyrosyl radicals and/or nitrogen dioxide. Thus, myeloperoxidase-dependent formation of 3-nitrotyrosine will be maximized at sites of inflammation when a cosubstrate, such as tyrosine is available, and the concentration of superoxide is low.

Table 8.1. Nitration of tyrosyl residues by stimulated neutrophils. Neutrophils (2×10^6 cells/ml) were incubated at 37°C in 10 m*M* phosphate-buffered saline (pH 7.4) containing 1 mg/ml glucose, 0.5 m*M* magnesium chloride, 1 m*M* calcium chloride, 100 μ*M* acetyl-Ser-Gln-Asn-tyr-Pro-Val-Val-NH_2, 1 m*M* nitrite, and one of 100 μ*M* tyrosine, 10 μg/ml superoxide dismutase, 1 m*M* methionine, 100 μ*M* azide, or 20 μg/ml catalase. Cells were stimulated with 100 ng/ml of 4β-phorbol-12-myristate-13-acetate for 1 h and products were then analysed by HPLC and quantified by comparison to the response for authentic 3-nitrotyrosine. The results are presented as means and ranges of duplicate determinations and are representative of three experiments with cells from different individuals [13]

Experimental condition	[Nitrated peptide] (μ*M*)
Control	0.87 + 0.16
+ Tyrosine	1.58 + 0.16
+ Superoxide dismutase	3.58 + 0.13
+ Methionine	1.41 + 0.16
+ Azide	0.03 + 0.03
+ Catalase	0.02 + 0.02

8.4 Hypochlorous Acid Production within Neutrophil Phagosomes

With our current understanding of redox transformations of myeloperoxidase, it is now possible to predict how the enzyme should function during bacterial killing. When neutrophils use myeloperoxidase to kill microorganisms, the action takes place inside a very narrow phagosomal space between the plasma membrane and the bacterial surface. Very little extracellular fluid is ingested with the bacterium, and the major constituents of the phagosome are the granule contents that empty into it. This creates an environment with a very high concentration of proteins, including myeloperoxidase, into which superoxide is released at a high rate. This environment is substantially different from the conditions normally employed to study the enzymology of myeloperoxidase. As described above, myeloperoxidase undergoes a complex array of reactions with different substrates, and its activity can be either inhibited or enhanced by superoxide. For these reasons, it is difficult to predict how well it will function in the phagosome, or what factors will be critical in determining the efficiency of hypochlorous acid production. Also because of accessibility problems, it is difficult to measure activity in the phagosome directly.

To try to understand the oxidative reactions occurring in the phagosome, we have developed a kinetic model of the reactions of myeloperoxidase. We have used the model to address the following questions:

- What concentrations of superoxide and hydrogen peroxide are achieved?
- Does HOCl account for the majority of the oxygen consumed during phagocytosis?
- Do reactions of superoxide with myeloperoxidase influence HOCl production?
- Is hydroxyl radical a significant secondary product of HOCl?

The model consists of more than 20 reactions and their rate constants and is described in detail elsewhere (Winterbourn et al., manuscript in preparation). The rate equations in the model describe the generation of superoxide, the interactions of superoxide, hydrogen peroxide and myeloperoxidase substrates with the redox intermediates of the enzyme (see Fig. 8.1), loss of hydrogen peroxide into the cytoplasm, and reactions of HOCl with superoxide and other potential targets. These equations were solved over time using the computer program, MATLAB. The main outputs were:

- steady-state superoxide and hydrogen peroxide concentrations
- efficiency of HOCl production from the oxygen consumed
- redox state of myeloperoxidase

Based on what we know about neutrophil oxidant production and phagosomal conditions, we made a number of assumptions about substrate concentrations, as shown in Table 8.2. The flux of superoxide was calculated on the basis that maximal oxygen consumption occurs on ingestion of 15–20 bacteria, and most of this occurs at the phagosomal membrane [15]. The phagosomal volume reflects the narrow space surrounding ingested bacteria as seen on electron micrographs [16]. Since superoxide formation occurs in this narrow film, it should be noted that although the flux is extraordinarily high when expressed as a concentration, it does not represent a large amount

Table 8.2. Best estimates of conditions in the phagosome

Phagosome volume	10^{-15} l
Superoxide generation rate	6 m*M*/s
Myeloperoxidase concentration	1.6 m*M*
Chloride concentration	0.1 *M*
pH	7.4
Other potential myeloperoxidase substrates present at	0.1 m*M*

of oxygen consumed, and diffusion is likely to prevent oxygen depletion from being a limiting factor. The myeloperoxidase concentration is based on three or four azurophilic granules (or about 5% of the total enzyme) being released per phagosome. The neutral pH is as measured during early stages of phagocytosis [17, 18] when most killing occurs. Little is known about other potential myeloperoxidase substrates such as tyrosine, tryptophan, nitrite, or reactive protein groups, but it is assumed that low concentrations of these diffuse in or are released from granules or the bacterium. As described above such compounds may react well or poorly with compound II. The effect of varying the rate constant for this reaction has been tested in different simulations.

A simulation using our best estimates of phagosomal conditions is shown in Fig. 8.4. There is a very high flux of superoxide, most of which dismutates to give hydrogen peroxide, to maintain a steady-state superoxide concentration of about 10 μM. The hydrogen peroxide concentration does not build up and remains well below that of superoxide. Predictions from the model are consistent with spectral evidence [8], that superoxide initially reacts with the myeloperoxidase to convert most of it to compound III. The remainder is present as native enzyme. In order to see significant peroxidase activity or HOCl generation, the compound III must turn over. Although we have proposed that this occurs via reaction with hydrogen peroxide [19], we have found that this mechanism is much too slow to give any significant hypochlorous acid production. For myeloperoxidase to continue to function after the first few seconds, a reaction between compound III and superoxide must be invoked. Such a reaction has been proposed [19], and our studies with purified myeloperoxidase provide further evidence for it. Myeloperoxidase is then able to handle the high flux of superoxide and hydrogen peroxide such that a majority of the oxygen consumed (about 60% under these conditions) is converted to hypochlorous acid.

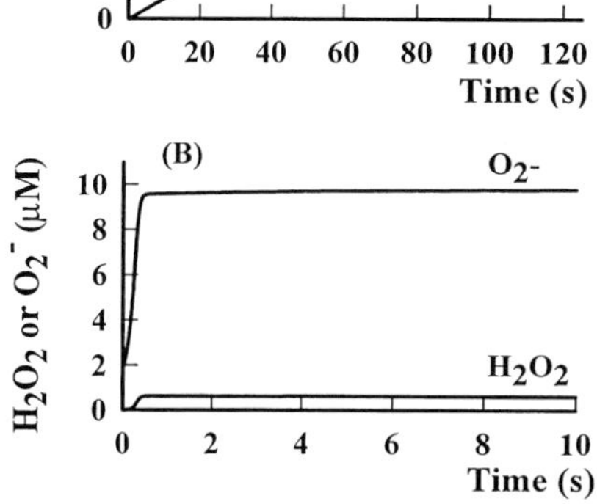

Fig. 8.4 A, B. Simulation of oxidant production in the phagosome, using a kinetic model incorporating the reactions shown in Fig. 8.1 and the conditions of Table 8.2

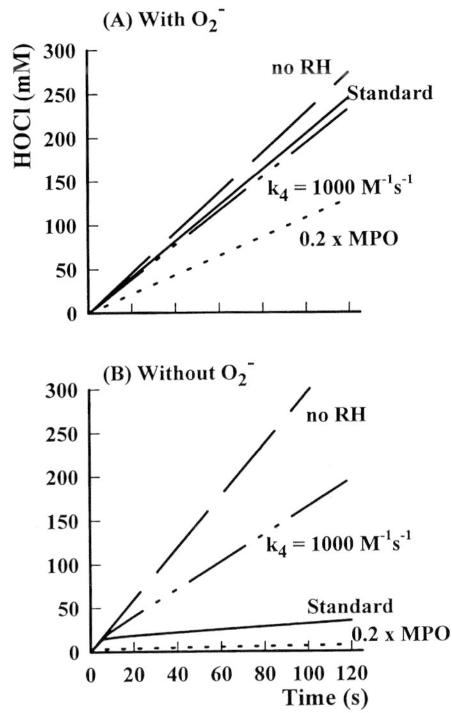

Fig. 8.5 A, B. Simulation of HOCl production and the effects of varying the myeloperoxidase concentration and the concentration and reactivity of an alternative myeloperoxidase substrate (electron donor) RH. **A** Standard conditions with 6 m*M*/s superoxide generated. **B** Hydrogen peroxide (3 m*M*/s) generated directly without production of superoxide. The maximum HOCl production in 1 min is 180 m*M* (— standard conditions as in Table 8.1 with a rate constant for reaction 4 of 100 $M^{-1}s^{-1}$, -- no electron donor present, - · · - · · standard conditions but k_4 1000 $M^{-1}s^{-1}$, -- standard conditions but with 20% the myeloperoxidase concentration)

We have examined the effects of varying different parameters on the efficiency of HOCl production, and some of these simulations are shown in Fig. 8.5. Removing or increasing the concentration of alternative myeloperoxidase substrates that can compete with chloride affects the efficiency of conversion of oxygen to HOCl, but only modestly. The myeloperoxidase concentration can be decreased fivefold with only a slight decrease in HOCl formation. Increasing the rate of cytoplasmic scavenging has the same effect (not shown). Likewise, the system is relatively insensitive to doubling the phagosomal volume or halving the superoxide generation rate. Thus, the system appears to be reasonably robust, with realistic variations in these variables not making large differences to the efficiency of hypochlorous acid formation.

We have also used the kinetic model to assess why it might be advantageous for the neutrophil to produce superoxide rather than hydrogen peroxide directly. When simulations were performed with a direct input of hydrogen peroxide rather than superoxide, we found that the reaction became much more sensitive to fluctuations in the different parameters (Fig. 8.5b). Under these conditions, complete conversion of oxygen to hypochlorous acid can be achieved, but only if no alternative myeloperoxidase substrates are present. Without superoxide to regenerate the native enzyme from compound II, myeloperoxidase becomes prone to inhibition by electron donors that readily reduce compound I but not compound II. Under our standard conditions of Table 8.1, the enzyme is converted to compound II and produces very little HOCl. The time it takes to achieve this, and the extent of inhibition, is highly dependent on the myeloperoxidase concentration as well as the concentration of the electron donor and its reactivity with compound II. It is likely that such compounds will be found in the phagosome, at least under some circumstances. So for the neutrophil to maintain its myeloperoxidase activity without stringent environmental requirements, there would be a clear advantage in generating superoxide.

We have also considered potential targets for HOCl within the phagosome, and the likelihood of its reacting with superoxide to produce hydroxyl radical. Although the rate constant for this reaction is high [20], there will be numerous other competing targets for HOCl. In particular, the high protein concentrations present initially in the phagosome as a result of degranulation will provide numerous highly reactive sites. We estimate thiols and methionine concentrations of 0.06 *M* and 0.3 *M* amines, compared with 10 μ*M* superoxide. Simulations allowing for this competition give less than 0.01% of the HOCl reacting with superoxide. Even if only a hundredth of the estimated reactive protein groups are present, this conversion increases only to 0.3%. We predict, therefore, that virtually all the reactive protein groups would need to be modified before there was significant hydroxyl radical production. Only if superoxide and HOCl are formed in a zone free of competing substrates such as protein [21], could hydroxyl radicals play any role in the phagosome.

8.5 Conclusion

1. Using our best estimates of conditions in the phagosome, kinetic modelling predicts that the neutrophil generates substantial amounts of HOCl, making it a plausible candidate for bacterial killing.
2. Intraphagosomal concentrations of hydrogen peroxide remain extremely low.
3. Interactions between superoxide and myeloperoxidase are important for enabling the enzyme to overcome inhibition by weak electron donors and limiting fluctuations in HOCl production under varying environmental conditions.
4. Hydroxyl radical production from HOCl and superoxide is severely limited by effective competition from proteins and other potential HOCl targets.

8.6 Summary

Myeloperoxidase is a heme enzyme of neutrophils and monocytes that plays a critical role in microbial killing. It reacts with a myriad of substrates and catalyses several different reactions in which it undergoes a complicated series of redox transformations. Its main function is thought to be production of hypochlorous acid from hydrogen peroxide and chloride. However, it also produces hypothiocyanite from thiocyanate, and catalyses the formation of free radicals from organic substrates such as tyrosine and urate. At physiological concentrations, tyrosine has little impact on production of hypochlorous acid. In contrast, tryptophan acts as a potent inhibitor of the enzyme by converting it to an inactive form. Tyrosine and superoxide overcome inhibition by reducing inactive myeloperoxidase back to the native enzyme. Superoxide reacts with most of the redox intermediates of myeloperoxidase and, depending on reaction conditions, either enhances or inhibits production of hypochlorous acid.

Nitrite is another physiologically relevant substrate for myeloperoxidase. Although poorly oxidized by the enzyme, cosubstrates such as tyrosine drive nitrite oxidation to nitrogen dioxide, which nitrates tyrosyl residues to form 3-nitrotyrosine. We have assimilated the known reactions of myeloperoxidase into a kinetic model to predict oxidant production within neutrophil phagosomes during microbial killing. Our model demonstrates that neutrophils should generate substantial amounts of hypochlorous acid, which makes it a plausible antimicrobial agent. In contrast, the steady-state concentrations of superoxide and hydrogen peroxide are extremely low. Reactions of superoxide with myeloperoxidase enable the enzyme to overcome inhibition by alternative substrates, such as tryptophan, and limit fluctuations in the production of hypochlorous acid under varying environmental conditions.

References

1. Kettle AJ, Winterbourn CC (1997) Myeloperoxidase: a key regulator of neutrophil oxidant production. Redox Rep 3: 3–15
2. Kettle AJ, Sangster DF, Gebicki JM, Winterbourn CC (1988) A pulse radiolysis investigation of the reactions of myeloperoxidase with superoxide and hydrogen peroxide. Biochim Biophys Acta 956: 58–62
3. Van Dalen CJ, Whitehouse M, Winterbourn CC, Kettle AJ (1997) Oxidation of thiocyanate by myeloperoxidase in the presence of chloride. Biochem J 327: 487–492
4. Kettle AJ, Winterbourn CC (1994) Assays for the chlorination activity of myeloperoxidase. Methods Enzymol 233: 502–512
5. Kettle AJ, Winterbourn CC (1992) Oxidation of hydroquinone by myeloperoxidase: mechanism of stimulation by benzoquinone. J Biol Chem 267: 8319–8324
6. Job D, Dunford HB (1976) Substituent effect on the oxidation of phenols and aromatic amines by horseradish peroxidase compound I. Eur J Biochem 66: 607–614
7. Kettle AJ, Winterbourn CC (1989) Influence of superoxide on myeloperoxidase kinetics measured with a hydrogen peroxide electrode. Biochem J 263: 823–828
8. Winterbourn CC, Garcia R, Segal AW (1985) Production of the superoxide adduct of myeloperoxidase (compound III) by stimulated neutrophils, and its reactivity with H_2O_2 and chloride. Biochem J 228: 583–592
9. Cuperus RA, Muijsers AO, Wever R (1986) The superoxidase activity of myeloperoxidase: formation of compound III. Biochim Biophys Acta 871: 78–84
10. Kettle AJ, Winterbourn CC (1994) Superoxide-dependent hydroxylation by myeloperoxidase. J Biol Chem 269: 17146–17151
11. Jiang Q, Hurst JK (1997) Relative chlorinating, nitrating, and oxidising capabilities of neutrophils determined with phagocytosable probes. J Biol Chem 272: 32767–32772
12. Eiserich JP, Hristova M, Cross CE, Jones AD, Freeman BA, Halliwell B, Vliet A van der (1998) Formation of nitric oxide-derived inflammatory oxidants by myeloperoxidase in neutrophils. Nature 391: 393–397
13. van Dalen CJ, Winterbourn CC, Senthilmohan R, Kettle AJ (2000) Nitrite as a substrate and inhibitor of myeloperoxidase. Implications for nitration and hypochlorous acid production at sites of inflammation. J Biol Chem. 275:11638–11644
14. Eiserich JP, Cross CE, Jones AD, Halliwell B, Vliet A van der (1996) Formation of nitrating and chlorinating species by reaction of nitrite with hypochlorous acid. J Biol Chem 271: 19199–19208
15. Hampton MB, Kettle AJ, Winterbourn CC (1996) The involvement of superoxide and myeloperoxidase in oxygen-dependent bacterial killing. Infect Immun 64: 3512–3517
16. Hampton MB, Kettle AJ, Winterbourn CC (1998) Inside the neutrophil phagosome: oxidants, myeloperoxidase and bacterial killing. Blood 92: 3007–3017
17. Segal AW, Geisow M, Garcia R, Harper A, Miller R (1981) The respiratory burst of phagocytic cells is associated with a rise in vacuolar pH. Nature 290: 406–409
18. Cech P, Lehrer RI (1984) Phagolysosomal pH of human neutrophils. Blood 63: 88–95
19. Kettle AJ, Winterbourn CC (1988) Superoxide modulates the activity of myeloperoxidase and optimizes the production of hypochlorous acid. Biochem J 252: 529–536
20. Long CA, Bielski BHJ (1980) Rate of reaction of superoxide radical with chloride containing species. J Phys Chem 84: 555–557
21. Saran M, Michel C, Bors W (1998) Radical functions in vivo: a critical review of current concepts and hypotheses. Z Naturforsch 53: 210–227

Acquisition and Use of Myeloperoxidase in the Microbicidal Activity of Macrophages

Laszlo Marodi, Christopher Tourney, Rita Kaposzta, Richard B. Johnston Jr., Nicole Moguilevsky

9.1 Introduction

Myeloperoxidase (MPO), a basic hemoprotein enzyme present in the primary granules of neutrophils and monocytes plays a critical role in the fungicidal activity of these cells [1, 2]. In vitro phagocytes genetically deficient in MPO fail to kill *Candida albicans* and patients with hereditary MPO deficiency have an increased susceptibility to invasive *C. albicans* infections [3].

We have found previously that monocytes, which are rich in MPO, kill *C. albicans* at a significantly higher degree than do macrophages, which are known to lack MPO, and that monocytes and MPO-dependent oxidants kill four fewer pathogenic *Candida* species more effectively than they killed *C. albicans* [2]. We have subsequently evaluated the ability of recombinant human MPO (rhMPO) to augment the candidacidal activity of resident macrophages and those treated with recombinant human granulocyte-macrophage colony stimulating factor (rhGM-CSF). We found that rhMPO increases the candidacidal capacity of macrophages with a more profound effect on rhGM-CSF-activated cells.

9.2 Materials and Methods

Mononuclear cells were isolated from heparinized venous blood of healthy adults. The washed suspension of mononuclear cells was resuspended in DMEM with 2 m*M* l-glutamine supplemented with penicillin, streptomycin, and 10% heat-inactivated autologous serum and adjusted to a final concentration of 2.5×10^6 cells/ml [4]. The cells were incubated in Teflon beakers (Savillex, Minnetonka, Minn.) at 37°C in an atmosphere containing 5% CO_2 for 5 days. At the concentrations indicated, rhGM-CSF was added to macrophages after 72 h of culture, and the treatment was maintained for an additional 48 h. rhMPO was purified from transfected Chinese hamster ovary cells as previously described [5].

To prepare stationary growth phase yeast, *C. albicans* (ATCC 18804) was inoculated into 100 ml Sabouraud, 2% dextrose broth (Difco Laboratories, Detroit, Mich.) and cultured for 2–3 days at 30°C with continuous rocking. The viability of the *Candida* remained >97% as determined by the exclusion of 0.01% methylene blue (Fisher Scientific, Pittsburgh, Pa.). Preopsonization of *Candida* was performed by incubation of 5×10^6 cells/ml in the presence of 5% normal pooled serum for 30 min at 37°C with rotation at 4 rpm, followed by centrifugation and washes in KRPD at 4°C [6].

To assess fungicidal activity, preopsonized *Candida* cells and mononuclear phagocytic cells in suspension were incubated at 37°C with rotation at 4 rpm [7]. At various time-points, aliquots of the incubation mixture were removed and diluted in water. Phagocytic cells were disrupted by freezing in liquid nitrogen and thawing in a waterbath. The percentage of *Candida* cells that had been killed was determined by colony counting [7].

The release of O_2^- from macrophages was quantitated as the superoxide dismutase-inhibitable reduction of ferricytochrome c (type 111; Sigma, St Louis, Mo.) [4]. Preopsonized *Candida* cells (5×10^6) were added to macrophages (5×10^6) in KRPD buffer with 80 p*M* cytochrome c, with or without 50 pg/ml superoxide dismutase [4]. The reaction volume was 1.5 ml. Incubation was at 37 °C with rotation at 4 rpm.

The results are expressed as means±SEM. The symbol *n* refers to the number of experiments, each done in duplicate or triplicate. Statistical significance was determined using Student's *t*-test.

9.3 Results and Discussion

We incubated macrophages and serum-opsonized *Candida* in the presence or absence of rhMPO and studied the candidacidal activity of macrophages (Table 9.1). Preincubation of rhMPO with macrophages or the addition of rhMPO to the phagocytic mixture at the time of initiation of incubation resulted in a dose-related and significant increase in the killing of *Candida*, a plateau being achieved at a concentration of 3.2 U/ml rhMPO. Equivalent results were achieved by preincubating macrophages with rhMPO for 30 min, then centrifuging and washing the cells before adding *Candida* (Table 9.1). These data suggest that the effect of rhMPO was achieved primarily through binding to and internalization by macrophages. Maximal augmentation of both O_2^- release and candidacidal activity was achieved with 200 U/ml rhGM-CSF, with minimal or no further increase being achieved by treatment of macrophages with 300 or 500 U/ml rhGM-CSF. Accordingly, we used 200 U/ml rhGM-CSF, which achieved maximal activation, to study the effect of rhMPO on candidal killing by activated cells. Macrophages were incubated with preopsonized *Candida* in the presence of various concentrations of rhMPO (Table 9.2). In these experiments, rhMPO was added before the *Candida* at the beginning of the incubation. In the absence of rhMPO, killing of *Candida* by rhGM-CSF-treated macrophages was significantly higher than that by resident cells ($P<0.05$). The effect of rhMPO on the capacity of rhGM-CSF-activated macrophages to kill *Candida* was dose-dependent, and the presence of 0.8 U/ml rhMPO or more resulted in significant increases in the killing capacity of macrophages (Table 9.2).

We compared also the effect of rhMPO on the candidacidal activity of monocytes, which contain MPO, and macrophages, which do not. The killing capacity of monocytes was not affected by the addition of 3.2 U/ml rhMPO (data not shown). In contrast, macrophages killed *Candida* at a significantly higher rate and extent in the presence of 3.2 U/ml rhMPO under the same experimental conditions ($P<0.01$ after both 60 min and 120 min of incubation). The results of these studies indicate that exogenous rhMPO can upmodulate macrophage candidacidal function in vitro, whereas

Table 9.1. Effect of rhMPO on killing of *C. albicans* by resident macrophages. Incubation of macrophages (5×10^6/ml) and preopsonized *C. albicans* (5×10^6/ml) was performed under slow rotation (4 rpm) for 60 min. Data are means ± SEM ($n > 5$). *$P < 0.05$ vs buffer control

Addition of rhMPO[a]	Concentration of rhMPO (U/ml)	*Candida* killed (% of inoculum)
None	–	20 ± 6
Concomitant	0.8	22 ± 6
Concomitant	1.6	28 ± 9
Concomitant	3.2	36 ± 6*
Preincubation	3.2	32 ± 7*
Concomitant	6.4	35 ± 8
Preincubation	6.4	33 ± 7*

[a] Under "Concomitant" conditions MPO was added immediately before *Candida*. In other experiments, as indicated, macrophages were preincubated with the concentration of rhMPO shown for 30 min at 37 °C with rotation at 4 rpm, then centrifuged and washed twice with buffer

Table 9.2. Effect of rhMPO on killing of *C. albicans* by rhGM-CSF-treated macrophages. Incubation of macrophages (5×10^6/ml) and preopsonized *C. albicans* (5×10^6/ml) was performed under slow rotation (4 rpm) for 60 min. Data are means ± SEM ($n = 6$). Statistical significance was determined by comparing the killing of *Candida* by rhGM-CSF-treated macrophages in the absence and in the presence of rhMPO. *$P < 0.05$, **$P < 0.01$

Treatment of macrophages[a]	Concentration of rhMPO (U/ml)	Candida killed (% of inoculum)[b]
None	–	17 ± 8
rhGM-CSF	–	29 ± 5
rhGM-CSF	0.4	33 ± 7
rhGM-CSF	0.8	44 ± 6*
rhGM-CSF	1.6	54 ± 4**
rhGM-CSF	3.2	59 ± 7**
rhGM-CSF	6.4	57 ± 6**

[a] After 72 h in culture, macrophages were preincubated with buffer or 200 U/ml rhGM-CSF for another 48 h

the fungicidal activity mediated by endogenous MPO in monocytes is not enhanced by additional exogenous MPO.

The extent of the rhMPO enhancement of *C. albicans* killing was higher in association with activated macrophages, which have an accelerated respiratory burst, than with resident cells, and this effect occurred at lower concentrations of rhMPO. These findings support the concept that human macrophage candidacidal activity depends upon products of oxidative metabolism [2, 4, 8].

9.4 Summary

The results of the studies reported here clearly suggest that exogenous rhMPO augments candidacidal capacity of both resident and cytokine-activated macrophages in vitro. Recent animal studies have shown partial protection against rickettsiosis of mice treated with rhMPO [9]. Administration of exogenous rhMPO to mice infected with *Cowdria ruminatum* increases survival significantly, and rhMPO conjugated with the Fc domain of human IgG1 is more effective than rhMPO [9]. These data suggest that the effect of rhMPO shown here in vitro might also be achieved by in vivo administration. Taken together, these encouraging in vitro observations and animal studies strongly support the need for further research into the possible clinical application of rhMPO, particularly for the treatment of disseminated candidiasis in patients with profound neutropenia.

Acknowledgements. This work was supported by grants from the European Commission (PECO No. CIPD CT 940303), and the National Science Foundation of Hungary (OTKA TO17100) to Dr. Marodi, and by the March of Dimes Birth Defects Foundation and a grant from the National Institutes of Health (AI 24782) to Dr. Johnston.

References

1. Lehrer RI (1969) Antifungal effects of peroxidase systems. J Bacteriol 99: 361
2. Marodi L, Forehand JR, Johnston RB Jr (1991) Mechanisms of host defense against *Candida* species II. Biochemical basis for killing of *Candida* by mononuclear phagocytes. J Immunol 146: 2790–2794
3. Nauseef WM (1988) Myeloperoxidase deficiency. Hematol Oncol Clin North Am 2: 135–147
4. Marodi L, Schreiber S, Anderson D, MacDermott RP, Korchak HM, Johnston RB Jr (1993) Enhancement of macrophage candidacidal activity by IFN-γ: Increased phagocytosis, killing, and calcium signal mediated by a decreased number of mannose receptors. J Clin Invest 91: 2596–2601
5. Moguilevsky N, Garcia-Quintana L, Jacquet A, Tournay C, Fabry L, Pirard R, Bollen A (1991) Structural and biological properties of human recombinant myeloperoxidase produced by Chinese hamster ovary cell lines. Eur J Biochem 197: 605–614
6. Marodi L, Korchak HM, Johnston RB Jr (1991) Mechanisms of host defense against *Candida* species 1. Phagocytosis by monocytes and monocyte-derived macrophages. J Immunol 146: 2783–2789
7. Marodi L, Kaposzta R, Campbell DE, Polin RA, Csongor J, Johnston RB Jr (1994) Candidacidal mechanisms in the human neonate: impaired IFN-γ activation in newborn infants. J Immunol 153: 5643–5649
8. Levitz SM, Diamond RD (1984) Killing of *Aspergillus fumigatus* spores and *Candida albicans* yeast phase by the iron-hydrogen peroxide-iodide cytotoxic system: comparison with the myeloperoxidase-hydrogen peroxide-halide system. Infect Immun 43: 1100–1102
9. Tournay C, Courtoy PJ, Marodi L, Totte P, Werenne J, Jacquet A, Garcia-Quintana L, Bollen A, Moguilevsky N (1996) Uptake of recombinant myeloperoxidase, free or fused to Fcγ, by macrophages enhances killing activity toward microorganisms. DNA Cell Biol 15: 617–624

Cooperative Interactions of the Peroxidase and Nitric Oxide Pathways in Inflammatory Oxidant Production

Jason P. Eiserich, Bruce A. Freeman, Carroll E. Cross, Albert van der Vliet

10.1 Introduction

Since its discovery as the endothelial-derived relaxing factor (EDRF) [1], nitric oxide (·NO) has been found to play a critical role in the function of nearly all organ systems and cellular responses [2]. In the past decade, a large body of evidence has emerged implicating ·NO both as a regulatory effector molecule of the innate immune response [3] and as a mediator of oxidative tissue injury associated with acute and chronic inflammation [4]. The production of ·NO is catalyzed by both constitutive and inducible forms of nitric oxide synthases (NOS) through the oxidation of the amino acid arginine [5]. It is the high-output inducible form of NOS (iNOS) which appears to be primarily responsible for the pathophysiological actions and host-defense functions attributed to ·NO [6]. The inducible ·NO pathway equips phagocytic cells, as well as other cell types capable of upregulating iNOS in response to pro-inflammatory stimuli, with cytotoxic activity against bacteria, fungi, viruses and tumor cells [3]. Although the ·NO pathway has endowed the host with a powerful means of protection against invading pathogens, it is not without consequence; that is, toxicity to other host cells and tissues is an inevitable process.

It is now well known that ·NO plays a central role in the pathogenesis of diverse inflammatory, infectious and degenerative disorders. However, the mechanisms underlying its effects on these various processes are only now emerging. Unlike many free radicals, ·NO displays highly selective reactivity, has a relatively long half-life and is capable of transmembrane diffusion due to its charge neutrality and small size. These attributes allow ·NO to function as a potent and selective biological mediator. Due to its relative hydrophobicity, ·NO readily partitions into lipid compartments and thereby increases its local concentration in cell membranes and lipoproteins and facilitates nitrogen dioxide ($\cdot NO_2$) formation by its reaction with O_2 [7]. Whereas direct reaction of ·NO with most biological molecules does not occur to an appreciable extent, its reaction with other free radicals (paramagnetic species) is extremely rapid. This is perhaps best illustrated by its facile reactions ($k = 1\text{–}3\times10^9\ M^{-1}s^{-1}$) with protein and amino acid radicals (e.g. tyrosine and tryptophan radicals) [8] and lipid peroxyl and alkoxyl radicals [9]. In addition, paramagnetic heme centers in proteins are targets of direct reactions of ·NO [10], which in many cases has been shown to regulate enzyme activity [11]. It is this selective reactivity which is thought to dictate the behavior and fate of ·NO in biological systems. In accordance with this theory, the toxicity of ·NO under inflammatory conditions has been proposed to be engendered by its nearly diffusion-limited reaction with superoxide anion ($O_2\cdot^-$), generated by the action of the respiratory burst oxidase of phagocytic cells, to produce the potent oxidant and nitrating agent peroxynitrite ($ONOO^-$, scheme 1) [12].

$$\cdot NO + O_2\cdot^- \rightarrow ONOO^- \qquad (1)$$

Since $ONOO^-$ is a short-lived species ($t_{1/2}$ about 1 s in aqueous media), it cannot be directly measured in cellular systems or in the human body. However, in vitro chemical studies have demonstrated that $ONOO^-$ reacts with the amino acid tyrosine to form 3-nitrotyrosine (scheme 2, NO_2Tyr) [13–15], a stable product that has been utilized as a "marker" or "footprint" of $ONOO^-$ production in vivo. Indeed, NO_2Tyr is significantly elevated in diverse human diseases and clinical disorders such as atherosclerosis, rheumatoid arthritis, acute and chronic lung injury, inflammatory bowel disease, organ rejection following transplantation, myocardial inflammation, neurodegeneration and tobacco smoking (reviewed in reference 16).

$$\text{Tyrosine} \rightarrow NO_2Tyr \qquad (2)$$

Metabolism of NO in the human body through a number of distinct pathways eventually leads to the formation of nitrite (NO_2^-) and nitrate (NO_3^-) as oxidative byproducts. Indeed, NO_2^- and NO_3^- have been utilized as "markers" of $\cdot NO$ production in cellular systems and biological fluids and tissues. In healthy human subjects, NO_2^- can be detected at levels of 0.5–3.6 μM in plasma, 15–20 μM in respiratory tract lining fluids, 30–120 μM in saliva, and 0.4–60 μM in gastric juice (reviewed in references 17 and 18). In addition, NO_2^- levels are markedly increased during inflammatory processes, reflecting increased $\cdot NO$ production, and in many cases parallel the number of phagocytes recruited into the tissue or extracellular fluid [19]. Increased levels of NO_2^- have been detected in synovial fluids of patients with rheumatoid arthritis (1–4 μM) [20], and in serum (about 35 μM) of human immunodeficiency virus-infected patients with interstitial pneumonia [21], dramatically higher than normal serum levels. Moreover, increased NO_2^- levels have been detected in condensed exhales from patients with asthma compared with those of healthy subjects [22].

Although NO_2^- is a major end-product of $\cdot NO$ metabolism, a large body of evidence reveals that it does not accumulate in vivo but is further metabolized to NO_3^-, implying that physiological pathways of NO_2^- oxidation are operative in vivo [23]. Early studies have illustrated that NO_2^- is oxidized to NO_3^- by compound I of catalase [24] and by oxyhemoglobin or oxymyoglobin [25, 26]. More recently, Klebanoff [27] has shown that NO_2^- can modulate the antimicrobial activity of myeloperoxidase (MPO), further supporting a physiological interaction between $\cdot NO$-derived species and peroxidases. However, the mechanistic basis for these observations has not been fully elucidated. Based upon these historical observations [23–27], the obligatory role of phagocytic peroxidases in inflammatory events [28] and the levels of NO_2^- attained in tissues and fluids under inflammation [20–22], we hypothesized that oxidation of NO_2^- by peroxidases may be an operative physiological pathway in host defense and tissue injury. In addition, due to the basic requirements for tyrosine nitration (i.e. addition of an NO_2 group into the ring of tyrosine), NO_2^- oxidation could facilitate a novel means of achieving tyrosine nitration in a manner independent of $ONOO^-$ formation (scheme 2).

10.2 Reactions Between Nitrite and Hypochlorous Acid

MPO is a heme protein located in the azurophilic granules of phagocytic cells that catalyzes chloride (Cl^-) oxidation by hydrogen peroxide (H_2O_2) to form hypochlorous acid (HOCl) (scheme 3) [29], a potent oxidizing agent with microbicidal activity [30].

$$H_2O_2 + H^+ + Cl^- \xrightarrow{\text{MPO}} HOCL + H_2O \qquad (3)$$

It has long been thought that the reaction of NO_2^- with HOCl results in the formation of nitrate (NO_3^-) (scheme 4) and it was thought to represent a classical example of an oxygen transfer reaction.

$$NO_2^- + HOCl \rightarrow H^+ + Cl^- + NO_3^- \qquad (4)$$

However, reevaluations of this reaction have suggested the formation of reactive intermediates known as nitryl chloride (Cl-NO_2) and/or chlorine nitrite (Cl-ONO) prior to NO_3^- generation [31]. The proposed reaction mechanism is schematically illustrated in Fig. 10.1a. Rather than oxygen atom transfer from HOCl to NO_2^-, the reaction proceeds via the transfer of a chlorinium ion (Cl^+) to NO_2^-. The reactive intermediates that are formed (Cl-NO_2 and Cl-ONO) then hydrolyze to form NO_3^- in the absence of organic substrates. These species are capable of nitrating, chlorinating

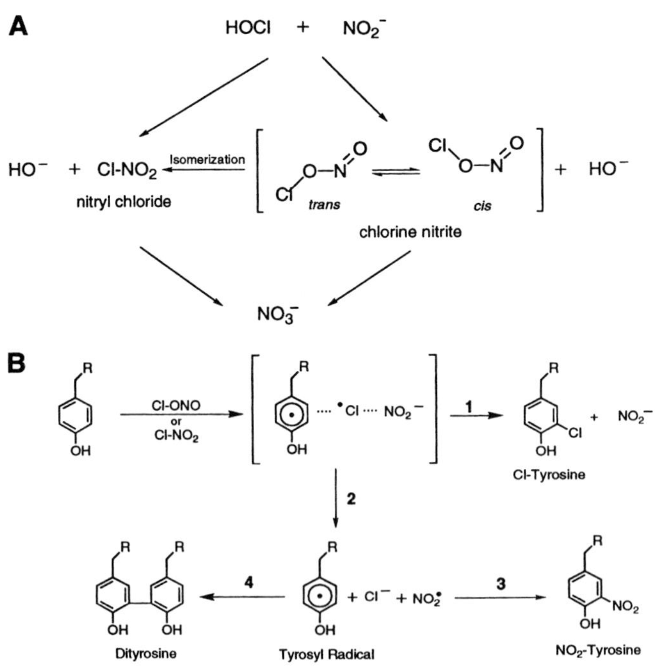

Fig. 10.1A,B. Proposed mechanisms of tyrosine chlorination, nitration and oxidation by reactive intermediates formed by the reaction of NO_2^- with HOCl. **A** NO_2^- reacts with HOCl via chloronium ion (Cl^+) transfer to yield nitryl chloride (Cl-NO_2) and/or chlorine nitrite (Cl-ONO) as intermediates in NO_3^- formation. **B** The reaction of Cl-NO_2 with tyrosine is proposed to proceed by electron transfer from tyrosine to $Cl^+NO_2^-$, resulting in an intermediate radical pair (tyrosyl radical-Cl·). Radical pair collapse (reaction 1) leads to formation of 3-chlorotyrosine (Cl-Tyrosine) and NO_2^- as major products. Dissociation of the complex from the solvent cage allows Cl· to oxidize NO_2^- to ·NO_2 (reaction 2), which can combine with simultaneously formed "free" tyrosyl radical (reaction 3) to yield 3-nitro-tyrosine (NO_2-Tyrosine). Dityrosine formation is achieved by the combination of two tyrosyl radicals (reaction 4). Adapted from reference 32

and oxidizing phenolic compounds such as the amino acid tyrosine and 4-hydroxyphenylacetic acid (HPA) [32, 33] as shown in Fig. 10.1b. Our studies on this reaction mechanism [32] have revealed that whereas chlorination reactions are mediated by direct electrophilic addition to the aromatic ring, a free radical mechanism appears to be operative in nitration reactions by NO_2Cl, most likely involving the combination of nitrogen dioxide ($\cdot NO_2$) and phenoxyl radicals ($ArO\cdot$). The pathophysiological relevance of this reaction pathway has since been illustrated by findings that NO_2^- can enhance HOCl-mediated lipid peroxidation and tyrosine nitration in human low-density lipoproteins [34]. Although not studied to date, hypobromous acid (HOBr) produced as a result of bromide (Br^-) oxidation by the eosinophil peroxidase (EPO) [35] may also react with NO_2^- to form nitryl bromide (NO_2Br), a species that could also catalyze tyrosine nitration and bromination. These reactions may become particularly important in pathologies such as asthma, where the activity of eosinophils and ·NO-dependent pathways are implicated [22].

10.3 Peroxidase-Catalyzed Nitrite Oxidation

Heme proteins such as catalase, metmyoglobin, methemoglobin, horseradish peroxidase (HRP), lactoperoxidase (LPO), MPO and EPO, when combined with H_2O_2, are capable of consuming NO_2^-, presumably forming NO_3^- [24–27]. In the case of HRP, it has been proposed [36] that oxidation of NO_2^- proceeds by sequential one-electron oxidation steps resulting in the formation of NO_3^- and regeneration of ferric HRP as shown in scheme 5.

$$\begin{aligned}
&\text{HRP} + H_2O_2 \rightarrow \text{HRP Compound I}\\
&\text{HRP Compound I} + NO_2^- \rightarrow \text{HRP Compound II} + \cdot NO_2\\
&\text{HRP Compound II} + NO_2^- \rightarrow \text{HRP} + \cdot NO_2\\
&2NO_2 + H_2O \rightarrow NO_2^- + NO_3^- + 2H^+\\
&2NO_2^- + H_2O_2 + H_2O \rightarrow NO_2^- + NO_3^- + H^+
\end{aligned} \tag{5}$$

In addition to plant peroxidases, recent investigations have illustrated that the peroxide-activated mammalian counterparts MPO and LPO can also promote oxidation of NO_2^- to intermediates that facilitate the nitration of aromatic substrates [17] such as tyrosine and HPA. As shown in Fig. 10.2a, these studies have revealed that MPO is a much more effective catalyst of NO_2^--dependent nitration reactions than is LPO. The absence of Cl^- in these reaction pathways reveals that HOCl is not necessarily an obligatory component in NO_2^- oxidation (as discussed above), supporting direct oxidation by MPO and LPO. At physiological concentrations of NO_2^- (1–50 μM) and chloride (Cl^-), MPO can effectively nitrate phenolic substrates (Fig. 10.2b) when exposed to fluxes of H_2O_2 (4.0 μM/min) [17]. In addition, NO_2^- causes enhanced MPO-mediated Cl^- and phenolic oxidation as demonstrated by increased Cl-HPA and Di-HPA.

Analogous to HRP-mediated NO_2^- oxidation, it is likely that MPO and LPO utilize both compound I and compound II to oxidize NO_2^-. Competition studies have demonstrated that NO_2^- oxidation by MPO and LPO occurs even in the presence of alternative anionic substrates (e.g. Cl^-, Br^-, SCN^-) [17], revealing that NO_2^- itself is a physiological substrate of mammalian peroxidases. In addition to small phenolic substrates, the peroxidase/H_2O_2/NO_2^- system promotes nitration of tyrosine residues

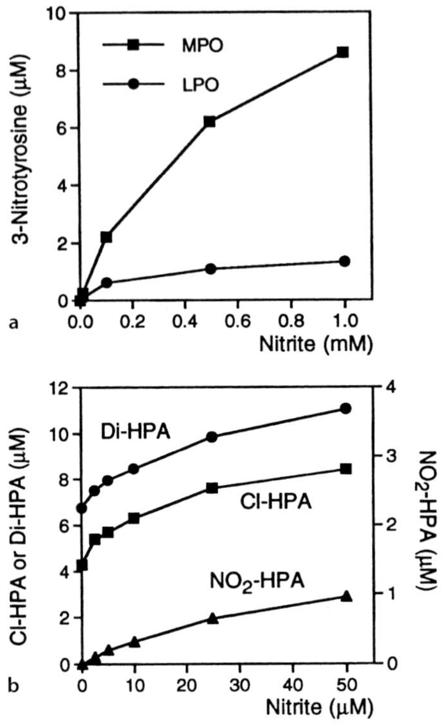

Fig. 10.2 a, b. Effect of NO_2^- on peroxidase-dependent tyrosine nitration, chlorination and oxidation. **a** Tyrosine nitration by MPO and LPO in the presence of NO_2^-. Tyrosine (1 m*M* in 50 m*M* phosphate buffer, pH 7.4) was incubated with the indicated concentrations of NO_2^- (0.1–1.0 m*M*), 280 µ*M* glucose, and 50 mU/ml glucose oxidase (generating about 4.0 µ*M* H_2O_2/min). The mixtures were incubated for 30 min at 37 °C in the presence of MPO or LPO (10 n*M*). **b** Effect of NO_2^- on phenolic modification by the MPO/H_2O_2/Cl^- system. HPA was incubated for 30 min at 37 °C in phosphate buffer (50 m*M*, pH 7.4) containing 150 m*M* Cl^-, MPO (10 n*M*) and the indicated concentrations of NO_2^-. H_2O_2 was enzymatically generated as described above (*Di-HPA* 3,3'-bis-HPA, *Cl-HPA* 3-chloro-HPA, *NO_2-HPA* 3-nitro-HPA). Adapted from reference 17

in purified proteins (e.g. bovine serum albumin [17]) as well as in more complex biological matrices (e.g. rat heart tissue homogenates [37]). Since proteins are not able to reach the restricted heme-containing active site of MPO and LPO, these latter observations suggest that the nitrating intermediates formed as a result of NO_2^- oxidation are highly diffusible. Studies performed by Sampson et al. [37] have revealed that tyrosine nitration reactions catalyzed by MPO/H_2O_2/NO_2^- remain operative in the presence of alternative oxidizable targets, such as thiols or ascorbic acid, further underscoring the contribution of these reactions pathways to protein tyrosine nitration observed in vivo [16].

10.4 Formation of Reactive Nitrogen Species by Activated Neutrophils

The physiological relevance of the aforementioned reaction pathways has been recently demonstrated in activated human PMN [18]. As shown in Fig. 10.3, PMN activated with phorbol myristate acetate (PMA) catalyzes the chlorination of phenolic substrates, confirming the results of studies reported by Domigan et al. [38] and Kettle [39]. Although no nitration reactions were observed under these conditions, inclusion of NO_2^- at physiologically relevant concentrations in the PMN suspension prior to activation resulted in a dose-dependent increase in the chlorination and nitration of HPA (Fig. 10.3). It is noteworthy that NO_3^- could not substitute for NO_2^- in

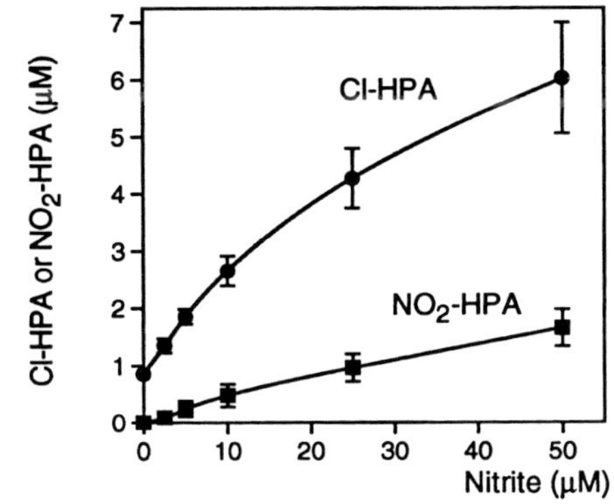

Fig. 10.3. Physiological concentrations of nitrite enhances neutrophil-mediated chlorination and nitration of phenolic substrates. Isolated human neutrophils (2×10^6/ml) were suspended in phosphate-buffered saline containing glucose (1 mg/ml) and HPA (1 m*M*), and incubated with PMA (50 ng/ml) for 60 min at 37 °C (*Cl-HPA* 3-chloro-HPA, *NO_2-HPA* 3-nitro-HPA). Adapted from reference 18

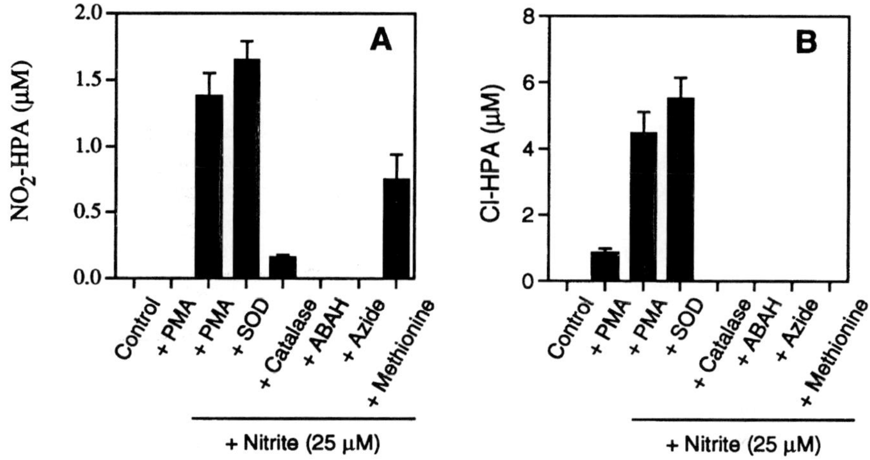

Fig. 10.4A,B. Nitrite-dependent nitration and chlorination of phenolic substrates is dependent upon H_2O_2 and active MPO. Incubation conditions were as detailed in Fig. 3 (*ABAH* 4-aminobenzoic acid hydrazide, *SOD* Cu,Zn-superoxide dismutase, *PMA* phorbol myristate acetate). Adapted from reference 18

these reaction pathways. As shown in Fig. 10.4, addition of Cu,Zn-superoxide dismutase (SOD) marginally enhanced, whereas catalase completely inhibited, the formation of Cl-HPA and NO_2-HPA, revealing a critical role of H_2O_2. Inhibitors of MPO, such as aminobenzoic acid hydrazide (ABAH) and azide, completely abolished nitration and chlorination reactions, and caused > 95% inhibition of Di-HPA, a characteristic product of MPO-mediated phenolic oxidation [40]. When methionine was added to scavenge HOCl, chlorination reactions were completely abolished whereas nitration yields were only partially decreased (about 50%), suggesting that oxidation of NO_2^- was occurring through both HOCl reaction and by direct interactions with H_2O_2-activated MPO. Using gas chromatography/mass spectrometry, ^{15}N-labeled NO_2^- has been found to be incorporated into phenolic substrates [18], unequivocally demonstrating the role of NO_2^- in PMN-mediated nitration reactions. We have also

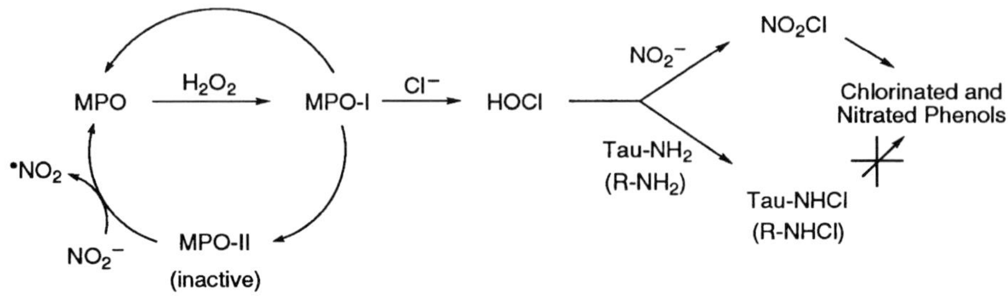

Fig. 10.5. Schematic illustration of the pathways by which nitrite enhances neutrophil-mediated chlorination of phenolic substrates. Adapted from reference 18

found that exposure of activated PMN to fluxes of $\cdot$NO also results in nitration reactions that are dependent upon H_2O_2 and active MPO, revealing that $ONOO^-$ is not responsible for phenolic nitration in this system [18].

The enhancement of MPO-mediated chlorination reactions we have observed suggests that NO_2^- may act as a catalyst in the peroxidase cycle. As summarized in Fig. 10.5, two possible mechanisms can account for the increases observed in PMN-dependent chlorination reactions. First, NO_2^- can compete with endogenous taurine for reactions with HOCl [18], forming NO_2Cl. This reaction generates a more selective chlorinating agent (NO_2Cl), and inhibits the formation of *N*-chlorotaurine, a species incapable of performing phenolic chlorination reactions. Second, similar to other reductants (e.g. ascorbate [41]), it has been proposed that NO_2^- may reduce the inactive MPO compound II, albeit at a slower rate than its reaction with compound I, to regenerate the native ferric enzyme and $\cdot NO_2$ [17]. Subsequent reactions of $\cdot NO_2$ with biological substrates (RH) regenerates NO_2^- (scheme 6), thus serving a catalytic role in MPO-mediated reactions.

$$\cdot NO_2 + RH \rightarrow NO_2^- + H^+ + R \qquad (6)$$

Based upon the data presented, we propose a novel function of peroxidases as "nitrotyrosine synthases". Since mammalian peroxidases exist as homodimeric enzymes, one-electron oxidation of both tyrosine (Tyr) and NO_2^- could theoretically occur either sequentially in a single subunit (Fig. 10.6a) or through the *cooperativity* of both subunits (Fig. 10.6b). In the case of the single active-site mechanism, it is likely that NO_2^- ($E°$ about 1.0 V [17]) is initially oxidized by MPO compound I forming $\cdot NO_2$ and compound II, which is then rapidly reduced by tyrosine ($E°$ about 0.85 V [17]) to regenerate ferric MPO and form Tyr$\cdot$. That is, a complete peroxidatic cycle is accomplished. Kettle et al. have recently provided support for this hypothesis [42]. They have observed that NO_2^- at physiological concentrations converts MPO compound I to compound II, the inactive form of MPO for Cl^- oxidation, indicating one-electron oxidation of NO_2^-. It is also possible that both subunits of MPO work in concert to simultaneously oxidize either substrate (Tyr and NO_2^-). Indeed, this may provide at least one fundamental explanation for the dimeric structure of mammalian peroxidases; that is, cooperative reactivity. In either case, the extremely rapid combination of $\cdot NO_2$ and Tyr$\cdot$ ($k = 3 \times 10^9\,M^{-1}s^{-1}$ [43]) generates the stable product NO_2Tyr which then can diffuse from the active site and impact on host cell function. Due to the

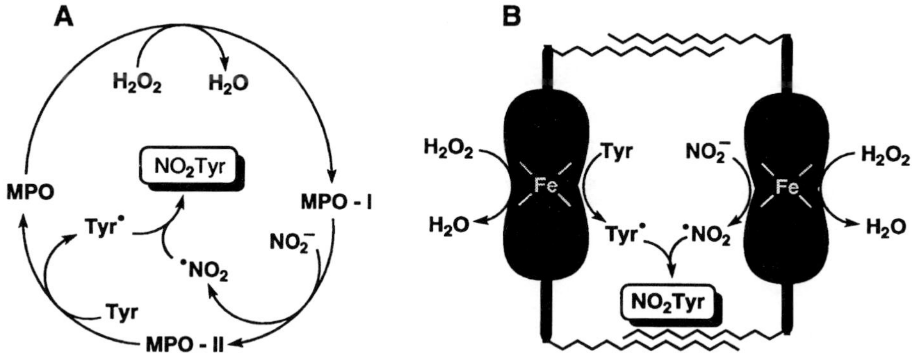

Fig. 10.6A,B. Proposed mechanisms for "nitrotyrosine synthase" activity of MPO. **A** Single active site model. NO_2^- reduces compound I of MPO to compound II and itself undergoes simultaneous one-electron oxidation forming $\cdot NO_2$. Compound II is then rapidly reduced to the native ferric MPO by tyrosine yielding Tyr·. Rapid combination of Tyr· and $\cdot NO_2$ forms 3-nitrotyrosine (NO_2Tyr) as a stable, non-radical, product that can diffuse from the active-site of MPO. **B** Concerted oxidation of NO_2^- and tyrosine through the cooperativity between both active sites in MPO

restricted active site of mammalian peroxidases, however, this model is limited to small, diffusible substrates (i.e. free tyrosine).

10.5 Physiological and Pathophysiological Implications

The discovery of the mammalian ·NO pathway has prompted a critical reevaluation of the biochemical and cellular mechanisms underlying host defense and tissue injury associated with phagocytic activity during inflammatory events. The studies described herein have confirmed previous suspicions and further elaborated on the interactive role between peroxidases and ·NO-derived species, namely NO_2^-. Although historically considered an innocuous byproduct of ·NO metabolism, NO_2^- now emerges as an important physiological player integrated into the myriad of functions executed by phagocyte peroxidases. The concurrent activation of both ·NO synthases and peroxidases suggests the possibility of novel mechanisms of phagocyte antimicrobial activity, only recently suggested by Klebanoff [27]. Formation of $\cdot NO_2$ and/or NO_2Cl by the $MPO/H_2O_2/Cl^-/NO_2^-$ system may enhance both the breadth and strength of the phagocyte host defense repertoire. First, reactions of $\cdot NO_2$ and NO_2Cl with biological macromolecules such as lipids, proteins and DNA of phagocytosed invading pathogens may result in direct cytotoxicity. Phagosomal colocalization of MPO, iNOS and nitrotyrosine in ingested bacteria [44] provides contributing support for the $MPO/H_2O_2/NO_2^-$ pathway in bactericidal actions of human PMN. Second interaction of the ·NO and peroxidase pathways leading to oxidative activation of phagocytic proteases may provide a complimentary and synergistic function to other host defense mechanisms of the phagocyte, as has been recently demonstrated for procollagenase of human PMN [45]. However, as NO_2^- has been shown to decrease the bactericidal activity of HOCl under in vitro conditions [27], an increase in host infection may surface during the coexpression of the

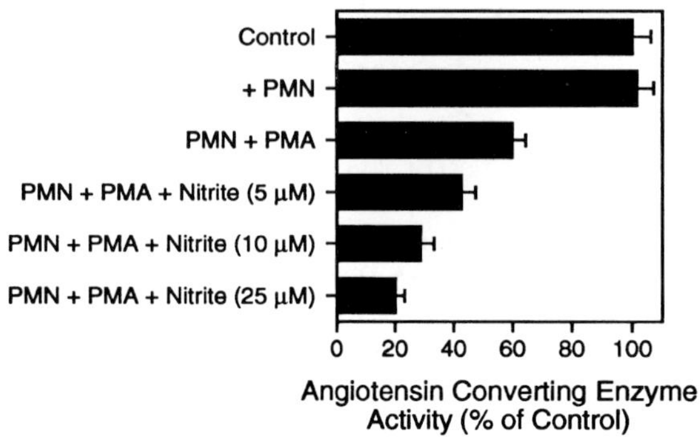

Fig. 10.7. Neutrophil-mediated inactivation of endothelial cell ACE activity is exacerbated by NO_2^-. Confluent bovine aortic endothelial cells (BAEC) were cocultured with isolated human PMN (5×10^5). Cocultures were activated with PMA (5 ng/ml) in the presence or absence of NO_2^- (5–25 μM) for 1.5 h at 37 °C. As shown above, non-activated PMN had no effect on ACE activity. However, the addition of PMA induced an approximate 40% loss of ACE activity. The addition of NO_2^- prior to PMA activation led to further increases in ACE inactivation. Adapted from reference 18

·NO and peroxidase pathways. Further studies elucidating these issues are clearly warranted.

The generation of reactive nitrogen species from peroxidase/NO_2^- interactions also provides insight into previously unrecognized mechanisms of oxidative tissue injury. Since inflammatory responses are causally linked to activation and adherence of PMN to target tissues, membrane proteins emerge as highly susceptible and well-poised targets of reactive nitrogen species produced from phagocytes. This concept is supported by observations that activated PMN utilize the MPO/H_2O_2 system to promote inactivation of the angiotensin-converting enzyme (ACE) [46], a protein located on the luminal aspect of endothelial cells responsible for processing and regulating vasoactive peptides [47]. We have recently demonstrated that PMN-dependent inhibition of ACE is further exacerbated by NO_2^- (Fig. 10.7) [18]. Moreover, the capacity of the peroxidase/H_2O_2/NO_2^- system to "synthesize" free nitrotyrosine has prompted investigations into the fate and toxicity of this modified amino acid. A recent study [48] has established that one-electron reduction of nitrotyrosine by NADH-cytochrome c reductase stimulates superoxide ($O_2{\cdot}^-$) production via intermediate formation of a nitrotyrosine anion radical, followed by electron transfer to O_2. Therefore, once formed by the MPO/H_2O_2/NO_2^-/tyrosine system, nitrotyrosine may actually promote oxidative stress by means of repetitive redox cycling. In addition, nitrotyrosine has the capacity to alter hemodynamic responses to α_1- and β-adrenoceptor agonists and angiotensin II [49, 50], suggesting that nitrotyrosine may be involved in the pathogenesis underlying hemodynamic disturbances associated with acute and chronic inflammatory conditions such as septic shock. In eukaryotic cells, NO_2Tyr induces microtubule dysfunction via its selective and irreversible incorporation into α-tubulin, a process catalyzed by the highly specific enzyme tubulin-tyrosine ligase (TTL; scheme 7) [51]. Collectively, these attributes of this unique amino acid derivative may provide novel mechanisms underlying the known toxicity of NO_2Tyr to mammalian cells [52] and its potential role as a mediator of tissue injury in vivo.

$$\alpha\text{-Tubulin-Glu-COOH} + NO_2\text{Tyr} \xrightarrow{\text{TTL}} \alpha\text{-Tubulin-Glu-}NO_2\text{Tyr-COOH} \qquad (7)$$

In addition to the formation of free NO_2Tyr, the MPO/H_2O_2/NO_2^- system promotes nitration of tyrosine residues in intact proteins [17, 37]. Over the past 30 years, protein chemists have used the chemical reagent tetranitromethane (TNM) as a selective means of inducing tyrosine nitration to explore the functional role of tyrosine residues in enzyme activity and protein structure. [53, 54]. The use of TNM has illustrated that nitration of key tyrosine residues is often associated with a loss of enzyme activity. Tyrosine nitration alters enzyme function through inducing substantial structural and electronic changes in tyrosine (scheme 2). First, the electron-withdrawing nature of the nitro functional group decreases the pK_a of tyrosine from 10 to about 7; that is, approximately 50% of the phenolic hydroxyl group in nitrotyrosine is deprotonated at physiological pH. Therefore, nitration introduces a negative charge in tyrosine that may either directly, or allosterically, alter the structure of proteins and hydrogen bonding interactions of tyrosine with substrates, neighboring amino acids or binding proteins. Second, nitration of tyrosine significantly increases its one-electron reduction potential (from about 0.85 V to about 1.05 V) (Eiserich et al., unpublished results). Third, the bulky nitro group induces steric perturbations in the structure of tyrosine. Indeed, nitration of tyrosine residues by reactive nitrogen species is known to compromise the structure and catalytic activity of numerous proteins in vitro such as manganese superoxide dismutase [55], surfactant protein A [56], prostacyclin synthase [57] and glutathione reductase [58]. It has also been demonstrated in vitro that nitration can inhibit kinase-dependent phosphorylation of tyrosine [59], suggesting that this covalent modification may disrupt critical cellular signaling pathways. Tyrosine nitration, then, may serve as a mechanism of cellular dysfunction, both to the invading pathogen as well as to host tissues.

Last, the discovery of a peroxidase-dependent NO_2^- oxidation pathway that leads to nitrotyrosine formation calls to into question the widespread use of this modified amino acid as a "specific" and "exclusive" marker of peroxynitrite formation in vivo. Several lines of evidence suggest that the MPO/H_2O_2/NO_2^- system may be an operative pathway of tyrosine nitration in vivo. First, NO_2Tyr is commonly detected in inflammatory diseases which are characterized by the infiltration and activation of leukocytes [16], and in many cases peroxidase activity is used to make this quantitative assessment [60]. Indeed, the levels of NO_2Tyr in lung tissues taken from patients with idiopathic pulmonary fibrosis parallel the number of peroxidase-positive inflammatory cells present in the tissue [61]. Second, NO_2Tyr residues detected in human pathologies such as atherosclerosis [62], cystic fibrosis [63] and multiple sclerosis [64] colocalize with high-level expression of MPO [65–67], the latter observation suggesting a role for NO_2^- and MPO in degenerative diseases of the central nervous system. Furthermore, detection of 3-chlorotyrosine, a specific marker of MPO-dependent chlorination reaction pathways has been demonstrated in human atherosclerotic lesions [68] and lends further credence to the MPO/NO_2^- pathway in human pathology.

With these issues raised, a strong case for MPO-dependent tyrosine nitration in vivo can be argued. Moreover, a recent report [69] has illustrated the ineffectiveness of $ONOO^-$ formed under physiological conditions, to nitrate tyrosine residues. Therefore, caution is advised when utilizing NO_2Tyr as a specific marker of peroxynitrite formation in diseased or injured tissues. Instead, it is our contention that NO_2Tyr be considered as a "footprint" left collectively by reactive nitrogen species formed in vivo. The presence of distinct, yet redundant, pathways for tyrosine nitration only further underscores the potential importance of this reaction pathway in host defense and tissue injury.

10.6
Future Directions and Remaining Enigmas

Despite the significant advancements that have recently emerged, further elucidation of the precise physiological and pathological implications of the peroxidase/NO_2^- pathways in vivo remains a challenge. Future studies employing transgenic MPO and iNOS knockout animal models may prove effective in resolving the controversial issues that remain. In addition, the use of MPO-deficient human phagocytes should prove to be an invaluable in vitro tool to further investigate the cooperative interactions of peroxidases and ·NO metabolites in host defense and tissue injury. Although it has been shown that iNOS-deficient mice display altered immune responses and increased susceptibility to viral and bacterial infections [70, 71], the precise reasons for this phenomenon remain poorly understood. Could it be that specific phagocyte-derived peroxidases are "starved" of crucial substrates such as NO_2^- arising from the ·NO pathway? Is it possible that the recently described ·NO system in plants [72] co-operates with peroxidases in providing a host defense mechanism [73] or even a physiological process of plant cell-wall maintenance? A better understanding of this cooperative relationship between peroxidases and ·NO-derived species will provide a means to elucidate novel functions of these distinct, yet overlapping biological pathways

10.7
Summary

Peroxidases and NO have independently emerged as key mediators of both host defense and oxidative tissue injury in plants and animals. As ·NO production and peroxidase activities are in many cases temporally and spatially colocalized during inflammatory events, we have proposed that these two distinct biochemical pathways converge and operate in a cooperative manner to produce reactive nitrogen intermediates capable of modifying biological molecules, most notably nitration of tyrosine. Through the use of chemical, enzymatic and cellular models we have demonstrated that NO_2^-, a major end-product of ·NO metabolism, is both a direct and indirect physiological substrate of peroxidase-dependent reactions. First, NO_2^- reacts with HOCl produced from the catalytic action of MPO to form the intermediate species NO_2Cl which is capable of nitrating, chlorinating and oxidizing phenolic substrates such as the amino acid tyrosine. Second, NO_2^- is directly oxidized by H_2O_2-activated peroxidases (i.e. MPO, LPO, HRP) to form $\cdot NO_2$. The concerted formation of Tyr· and $\cdot NO_2$ by MPO using H_2O_2, tyrosine and NO_2^- as substrates yields the modified amino acid 3-nitrotyrosine, a pathway that has led us to propose a novel function of peroxidases as "nitrotyrosine synthetases". Activated human polymorphonuclear leukocytes (neutrophils, PMN) are capable of converting physiological concentrations of NO_2^- into NO_2Cl and $\cdot NO_2$ through MPO-dependent pathways. These species can subsequently impact deleteriously on host cells, underscoring the relevance of these reacti ons in inflammatory processes. Moreover, nitration of critical tyrosine residues through these reactions can compromise the function of enzymes, thus leading to cellular dysfunction and tissue injury. Collectively, these observations provide a unifying concept of cooperativity between the peroxidase and ·NO pathways, and reveal a previously unrecognized mechanism underlying nitrotyrosine formation observed in diseased and inflamed tissues.

References

1. Furchgott RF, Zawadski JB (1980) The obligatory role of endothelial cells in the relaxation of arterial smooth muscle by acetylcholine. Nature 288: 373–376
2. Nathan C (1992) Nitric oxide as a secretory product of mammalian cells. FASEB J 6: 3051–3064
3. Fang FC (1997) Mechanisms of nitric oxide-related antimicrobial activity. J Clin Invest 99: 2818–2825
4. Kröncke KD, Fehsel K, Kolb-Bachofen V (1998) Inducible nitric oxide synthase in human diseases. Clin Exp Immunol 113: 147–156
5. Michel T, Feron O (1997) Nitric oxide synthases: which, where, how and why? J Clin Invest 100: 2146–2152
6. Nathan C (1997) Inducible nitric oxide synthase: what difference does it make? J Clin Invest 100: 2417–2423
7. Liu X, Miller MS, Joshi MS, Thomas DD, Lancaster JR Jr (1998) Accelerated reaction of nitric oxide with O_2 within the hydrophobic interior of biological membranes. Proc Natl Acad Sci U S A 95: 2175–2179
8. Eiserich JP, Butler J, Vliet A van der, Cross CE, Halliwell B (1995) Nitric oxide rapidly scavenges tyrosine and tryptophan radicals. Biochem J 310: 745–749
9. Rubbo H, Radi R, Trujillo M, Telleri R, Kalyanaraman B, Barnes S, Kirk M, Freeman BA (1994) Nitric oxide regulation of superoxide and peroxynitrite-dependent lipid peroxidation. Formation of novel nitrogen-containing oxidized lipid derivatives. J Biol Chem 269: 26066–26075
10. Radi R (1996) Nitric oxide reactions with metalloproteins. Chem Res Toxicol 9: 828–835
11. Ignarro LJ (1992) Heme-dependent activation of cytosolic guanylyl cyclase by nitric oxide: a widespread signal transduction mechanism. Biochem Soc Trans 20: 465–469
12. Beckman JS (1996) Oxidative damage and tyrosine nitration from peroxynitrite. Chem Res Toxicol 9: 836–844
13. Beckman JS, Ischiropoulos H, Zhu L, Woerd M van der, Smith C, Chen J, Harrison J, Martin JC, Tsai M (1992) Kinetics of superoxide dismutase- and iron-catalyzed nitration of phenolics by peroxynitrite. Arch Biochem Biophys 298: 438–445
14. Ischiropoulos H, Zhu L, Chen J, Tsai M, Martin JC, Smith CD, Beckman JS (1992) Peroxynitrite-mediated tyrosine nitration catalyzed by superoxide dismutase. Arch Biochem Biophys 298: 431–437
15. Vliet A van der, Eiserich JP, O'Neill CA, Halliwell B, Cross CE (1995) Tyrosine modification by reactive nitrogen species: a closer look. Arch Biochem Biophys 319: 341–349
16. Ischiropoulos H (1998) Biological tyrosine nitration: a pathophysiological function of nitric oxide and reactive oxygen species. Arch Biochem Biophys 356: 1–11
17. Vliet A van der, Eiserich JP, Halliwell B, Cross CE (1997) Formation of reactive nitrogen species during peroxidase-catalyzed oxidation of nitrite. A potential additional mechanism of nitric oxide-dependent toxicity. J Biol Chem 272: 7617–7625
18. Eiserich JP, Hristova M, Cross CE, Jones AD, Freeman BA, Halliwell B, Vliet A van der (1998) Formation of nitric oxide-derived inflammatory oxidants by myeloperoxidase in neutrophils. Nature 391: 393–397
19. Gaston B, Reilly J, Drazen JM, Fackler J, Ramdey P, Arnelle D, Mullins ME, Sugarbaker DJ, Chee C, Singel DJ, Loscalzo J, Stamler JS (1993) Endogenous nitrogen oxides and bronchodilator *S*-nitrosothiols in human airways Proc Natl Acad Sci U S A 90: 10957–10961
20. Farrell AJ, Blake DR, Palmer RMJ, Moncada S (1992) Increased concentrations of nitrite in synovial fluid and serum samples suggest increased nitric oxide synthesis in rheumatic diseases. Ann Rheum Dis 51: 1219–1222
21 Torre D, Ferrario G, Speranza F, Orani A, Fiori GP, Zeroli C (1996) Serum concentrations of nitrite in patients with HIV-1 infection. J Clin Pathol 49: 574–576
22. Hunt J, Byrns RE, Ignarro LJ, Gaston B (1995) Condensed expirate nitrite as a home marker for acute asthma. Lancet 346: 1235–1236

23. Parks NJ, Krohn KA, Mathis CA, Chasko JH, Geiger KR, Gregor ME, Peek NF (1981) Nitrogen-13-labeled nitrite and nitrate: distribution and metabolism after intratracheal administration. Science 212: 58–61
24. Heppel LA, Porterfield VT (1949) Metabolism of inorganic nitrite and nitrate esters. 1. The coupled oxidation of nitrite by peroxide-forming systems and catalase. J Biol Chem 178: 549–556
25. Keilin D, Hartree EF (1955) Catalase, peroxidase and metmyoglobin as catalysts of coupled peroxidatic reactions Biochem J 60: 310–325
26. Doyle MP, Herman JG, Dykstra RL (1985) Autocatalytic oxidation of hemoglobin induced by nitrite: activation and chemical inhibition. J Free Radic Biol Med 1: 145–153
27. Klebanoff SJ (1993) Reactive nitrogen intermediates and antimicrobial activity: role of nitrite. Free Radic Biol Med 14: 351–360
28. Weiss SJ (1989) Tissue destruction by neutrophils. New Engl J Med 320: 365–376
29. Klebanoff SJ (1991) Myeloperoxidase: occurrence and biological function. In: Everse J, Everse KE, Grisham MB (eds) Peroxidases in chemistry and biology. CRC, Boca Raton
30. Klebanoff SJ, Clark RA (1978) The neutrophil: function and clinical disorders. North Holland Publishing, Amsterdam
31. Johnson DW, Margerum DW (1991) Non-metal redox kinetics. A reexamination of the mechanism of the reaction between hypochlorite and nitrite ions. Inorg Chem 30: 4845–4851
32. Eiserich JP, Cross CE, Jones AD, Halliwell B, Vliet A van der (1996) Formation of nitrating and chlorinating species by reaction of nitrite with hypochlorous acid. A novel mechanism for nitric oxide-mediated protein modification. J Biol Chem 271: 19199–19208
33. Kono Y (1995) The production of nitrating species by the reaction between nitrite and hypochlorous acid. Biochem Mol Biol Int 36: 275–283
34. Panasenko OM, Briviba K, Klotz L-O, Sies H (1997) Oxidative modification and nitration of human low-density lipoproteins by the reaction of hypochlorous acid with nitrite. Arch Biochem Biophys 343: 254–259
35. Thomas EL, Bozeman PM, Jefferson MM, King CC (1995) Oxidation of bromide by the human leukocyte enzymes myeloperoxidase and eosinophil peroxidase. Formation of bromamines. J Biol Chem 270: 2906–2913
36. Roman R, Dunford HB (1973) Studies on horseradish peroxidase. XII. A kinetic study of the oxidation of sulfite and nitrite by compounds I and II. Can J Chem 51: 588–596
37. Sampson JB, Ye YZ, Rosen H, Beckman JS (1998) Myeloperoxidase and horseradish peroxidase catalyze tyrosine nitration in proteins from nitrite and hydrogen peroxide. Arch Biochem Biophys 356: 207–213
38. Domigan NM, Charlton TS, Duncan MW, Winterbourn CC, Kettle AJ (1995) Chlorination of tyrosyl residues in peptides by myeloperoxidase and human neutrophils. J Biol Chem 270: 16542–16548
39. Kettle AJ (1996) Neutrophils convert tyrosyl residues in albumin to chlorotyrosine. FEBS Lett 379: 103–106
40. Heinecke JW, Li W, Daehnke HL, Goldstein JA (1993) Dityrosine, a specific marker of oxidation, is synthesized by the myeloperoxidase-hydrogen peroxide system of human neutrophils and macrophages. J Biol Chem 268: 4069–4077
41. Marquez LA, Dunford HB, Van Wart H (1990) Kinetic studies on the reaction of compound II of myeloperoxidase with ascorbic acid. Role of ascorbic acid in myeloperoxidase function. J Biol Chem 265: 5666–5670
42. Kettle AJ, Dalen CJ van, Winterbourn CC (1997) Peroxynitrite and myeloperoxidase leave the same footprint in protein nitration. Redox Rep 3: 257–258
43. Prütz WA, Mönig H, Butler J, Land EJ (1985) Reactions of nitrogen dioxide in aqueous model systems. Oxidation of tyrosine units in peptides and proteins. Arch Biochem Biophys 243: 125–134
44. Evans TJ, Buttery LD, Carpenter A, Springall DR, Polak JM, Cohen J (1996) Cytokine-treated human neutrophils contain inducible nitric oxide synthase that produces nitration of ingested bacteria. Proc Natl Acad Sci U S A 93: 9553–9558
45. Okamoto T, Akaike T, Nagano T, Miyajima S, Suga M, Ando M, Ichimori K, Maeda H (1997) Activation of human neutrophil procollagenase by nitrogen dioxide and peroxynitrite: a novel mechanism for procollagenase activation involving nitric oxide. Arch Biochem Biophys 342: 261–274
46. Chen X, Catravas JD (1993) Neutrophil-mediated endothelial angiotensin-converting enzyme dysfunction. Role of oxygen-derived free radicals. Am J Physiol 265: L243–L249
47. Corvol P, Williams TA, Soubrier F (1995) Peptidyl dipeptidase A: angiotensin I-converting enzyme. Methods Enzymol 248: 283–305
48. Krainev AG, Williams TD, Begelow DJ (1998) Enzymatic reduction of 3-nitrotyrosine generates superoxide. Chem Res Toxicol 11: 495–502
49. Kooy NW, Lewis SJ (1996) The peroxynitrite product 3-nitro-l-tyrosine attenuates the hemodynamic responses to angiotensin II in vivo. Eur J Pharmacol 315: 165–170

50. Kooy NW, Lewis SJ (1996) Nitrotyrosine attenuates the hemodynamic effects of adrenoceptor agonists in vivo: relevance to the pathophysiology of peroxynitrite. Eur J Pharmacol 310: 155–161
51. Eiserich JP, Estévez AG, Bamberg T, Ye YZ, Beckman JS, Freeman BA (1999) Microtubule dysfunction by selective post-translational nitrotyrosination of α-tubulin. A nitric oxide-dependent mechanism of cellular injury. Proc Natl Acad Sci U S A 96: 6365–6370
52. Riccardi VM, Maragos VA (1980) The pathophysiology of neurofibromatosis. I. Resistance in vitro to 3-nitrotyrosine as an expression of the mutation. In Vitro 16: 706–714
53. Sokolovsky M, Riordan JF, Vallee BL (1966) Tetranitromethane: a reagent for the nitration of tyrosyl residues in proteins. Biochemistry 5: 3582–3589
54. Riordan JF, Sokolovsky M, Vallee BL (1967) Environmentally sensitive tyrosyl residues. Nitration with tetranitromethane. Biochemistry 6: 358–361
55. MacMillan-Crow LA, Crow JP, Thompson JA (1998) Peroxynitrite-mediated inactivation of manganese superoxide dismutase involves nitration and oxidation of critical tyrosine residues. Biochemistry 37: 1613–1622
56. Zhu S, Haddad IY, Matalon S (1996) Nitration of surfactant protein A (SP-A) tyrosine residues results in decreased mannose binding ability. Arch Biochem Biophys 333: 282–290
57. Zhou M, Martin C, Ullrich V (1997) Tyrosine nitration as a mechanism of selective inactivation of prostacyclin synthase by peroxynitrite. Biol Chem 378: 707–713
58. Francescutti D, Baldwin J, Lee L, Mutus B (1996) Peroxynitrite modification of glutathione reductase: modeling studies and kinetic evidence suggest the modification of tyrosines at the glutathione disulfide binding site. Protein Eng 9: 189–194
59. Kong S-K, Yim MB, Stadtman ER, Chock PB (1996) Peroxynitrite disables the tyrosine phosphorylation regulatory mechanism: lymphocyte-specific tyrosine kinase fails to phosphorylate nitrated cdc2(6–20)NH_2 peptide. Proc Natl Acad Sci U S A 93: 3377–3382
60. Schneider T, Issekutz AC (1996) Quantitation of eosinophil and neutrophil infiltration into rat lung by specific assays for eosinophil peroxidase and myeloperoxidase. Application in a Brown Norway rat model of allergic pulmonary inflammation. J Immunol Methods 198: 1–14
61. Saleh D, Barnes PJ, Giaid A (1997) Increased production of the potent oxidant peroxynitrite in the lungs of patients with idiopathic pulmonary fibrosis. Am J Respir Crit Care Med 155: 1763–1769
62. Beckman JS, Ye Y, Anderson PG, Chen Accavitti, MA Tarpey, MM, White CR (1994) Extensive nitration of protein tyrosines in human atherosclerosis detected by immunohistochemistry. Biol Chem Hoppe-Seyler 375: 81–88
63. Vliet A van der, Eiserich JP, Marelich GP, Halliwell B, Cross CE (1997) Oxidative stress in cystic fibrosis. Does it occur and does it matter? Adv Pharmacol 38: 491–513
64. Cross AH, Manning PT, Keeling RM, Schmidt RE, Misko TP (1998) Peroxynitrite formation within the central nervous system in active multiple sclerosis. J Neuroimmunol 88: 45–56
65. Daugherty A, Dunn JL, Rateri DL, Heinecke JW (1994) Myeloperoxidase, a catalyst for lipoprotein oxidation, is expressed in human atherosclerotic lesions. J Clin Invest 94: 437–444
66. Mohammed JR, Mohammed BS, Pawluk LJ, Bucci DM, Baker NR, Davis WB (1988) Purification and cytotoxic potential of myeloperoxidase in cystic fibrosis sputum. J Lab Clin Med 112: 711–720
67. Nagra RM, Becher B, Tortellotte WW, Antel JP, Gold D, Paladino T, Smith RA, Nelson JR, Reynolds WF (1997) Immunohistochemical and genetic evidence of myeloperoxidase involvement in multiple sclerosis. J Neuroimmunol 78: 97–107
68. Hazen SL, Heinecke JW (1997) 3-Chlorotyrosine, a specific marker of myeloperoxidase-catalyzed oxidation, is markedly elevated in low density lipoprotein isolated from human atherosclerotic intima. J Clin Invest 99: 2075–2081
69. Pfeiffer S, Mayer B (1998) Lack of tyrosine nitration by peroxynitrite generated at physiological pH. J Biol Chem 273: 27280–27285
70. MacMicking JD, Nathan C, Hom G, Chartrain N, Trumbauer M, Stevens K, Xie Q-W, Sokol K, Fletcher DS, Hutchinson N, Chen H, Mudgett JS (1995) Altered responses to bacterial infection and endotoxic shock in mice lacking inducible nitric oxide synthase. Cell 81: 641–650
71. Wei X-Q, Charles IG, Smith A, Ure J, Feng G-J, Huang F-P, Xu D, Muller W, Moncada S, Liew FY (1995) Altered immune responses in mice lacking inducible nitric oxide synthase. Nature 375: 408–411
72. Delledonne M, Xia Y, Dixon RA, Lamb C (1998) Nitric oxide functions as a signal in plant disease resistance. Nature 394: 585–588
73. Bolwell GP, Butt VS, Davies DR, Zimmerlin A (1995) The origin of the oxidative burst in plants. Free Radic Res 23: 517–532

The Role of the Monoclonal Antibody Anti-Myeloperoxidase (anti-MPO) in the Diagnosis and Classification of Acute Leukaemias

Estella Matutes

11.1 Introduction

Myeloperoxidase (MPO) serves as the hallmark for the myeloid lineage and is regarded as the most specific cytochemical marker for the diagnosis and definition of acute myeloid leukaemias (AML) [1]. Its cytochemical expression has been routinely investigated by light microscopy and less often by transmission electron microscopy (TEM), although the latter method has been shown to be more sensitive than light microscopy for detecting the enzyme activity [2].

Over the last decade, the availability of polyclonal and monoclonal antibodies (mAb) that detect the α-chain of mature MPO and the proenzyme form has permitted investigation of the presence of this enzyme by immunophenotypic techniques. Our experience and that of others has already shown that immunophenotyping is more sensitive than light microscopic cytochemistry for demonstrating MPO activity in undifferentiated blast cells from acute leukaemias, although only a small series of cases have been documented in the literature [3–6].

We describe here findings with the anti-MPO mAb in 74 patients with acute leukaemia difficult to classify by light microscopy and the majority of them by immunological markers. These comprised 17 cases of poorly differentiated AML, designated AML-Mo by the French-American-British (FAB) group and in which, by definition, blasts are MPO-negative by light microscopy [7]. In addition we analysed 57 cases of mixed lineage or biphenotypic acute leukaemia (BAL) in which the blasts are morphologically undifferentiated or exhibit myeloid features by light microscopy but because they coexpress myeloid and lymphoid antigens are difficult to classify as acute lymphoblastic leukaemias (ALL) or AML [8].

11.2 Material and Methods

11.2.1 Patients

Of the 74 patients studied, 17 were classified as AML-Mo and 57 as BAL according to the European group for the immunological classification of leukaemias (EGIL) [9]. Cells from all 17 AML-Mo were poorly differentiated blasts, negative by light microscopy for the cytochemical reactions Sudan black B (SBB), MPO and non-specific esterase (ANAE), expressed one or more myeloid antigens, and were negative with B and T lymphoid specific markers. Of the 57 cases of BAL, 20 had a “lymphoblastic” morphology, 31 showed features of myeloid differentiation, by morphology and/or cytochemistry. In 6 cases, slides were not available for review. Immunophenotyping showed that 36 cases had a myeloid and B lymphoid phenotype, 17 a myeloid and T lymphoid phenotype, 2 a B and T lymphoid phenotype and the remaining 2 showed trilineage differentiation. Negative and positive controls for anti-MPO included cases of ALL and AML and normal bone marrow myeloid precursors.

11.2.2 Immunostaining with Anti-MPO

Peripheral blood and/or bone marrow mononuclear cells were isolated on a density gradient centrifugation, washed in Hank's balanced salt solution, pelleted onto slides for immunocytochemistry staining and to assess the proportion of blasts. Immunostaining with the anti-MPO mAb (CLB-MPO) was performed either by immunocytochemistry by applying the immunoalkaline phosphatase anti-alkaline phosphatase (APAAP) technique as previously described [3] or by flow cytometry on fixed and permeabilized cells gating in the blast population.

A sample was judged anti-MPO-positive when over 5% of blasts showed staining with the antibody or when over 20% of cells expressed the marker by flow cytometry.

11.2.3 Transmission Electron Microscopy (TEM) Cytochemistry

MPO activity was investigated by three methods [2, 6]: (1) after fixation according to Graham and Karnovsky's method; (2) using unfixed cells according to Roel's method and (3) after a short fixation with glutaraldehyde in Gey's buffer. The latter two methods appear to be equally sensitive but more sensitive than Graham and Karnovsky's technique.

For the immunogold method, the cells were first incubated with the anti-lymphoid or anti-myeloid mAb, followed by a second incubation with a gold-conjugated goat antimouse immunoglobulin. When the immunological reaction was complete, the peroxidase reaction was performed according to the methods described above.

11.3 Results

11.3.1 Reactivity with the Anti-MPO mAb

A variable proportion of blasts (15–70%) from 8 out of 17 (47%) AML-Mo expressed anti-MPO (Table 11.1). The reactivity was seen as a red cytoplasmic deposit in the cytoplasm by the APAAP technique and/or as a peak above the negative control by flow cytometry (Fig. 11.1). Two other myeloid-associated markers, CD13 and CD33, were positive in 76% and 58% of cases, respectively whereas cells from the majority of cases reacted with markers expressed in early haemopoietic cells. A third of the cases were TdT-positive and close to half CD7-positive (Table 11.1). Cells from all AML-Mo

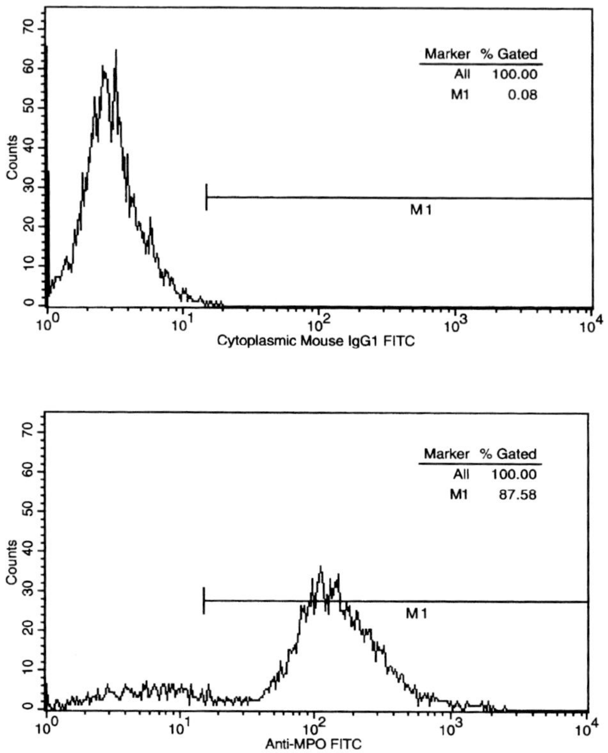

Fig. 11.1. Flow cytometry chart on a case of AML-M0 showing a high proportion of blasts expressing anti-MPO

Table 11.1. Anti-MPO mAb in minimally differentiated AML (AML-M0). All cases were MPO-negative by light microscopy and negative with CD2, cytCD3, CD79a, CD19 and CD22. TdT was positive in 30% and CD7 in 50%

mAb	Positive cases Number	%
anti-MPO	8/17	47
CD13	13/17	76
CD33	10/17	58

Table 11.2. Anti-MPO mAb in BAL. Of 57 cases, 20 had an ALL morphology and cytochemistry (MPO-negative by light microscopy) and 31 an AML morphology and cytochemistry. All cases coexpressed lymphoid and myeloid markers with scores >2 for each lineage

mAb	Positive cases Number	%
anti-MPO	53/57	93
CD13	51/57	89
CD33	36/57	63
CD14	4/20	20

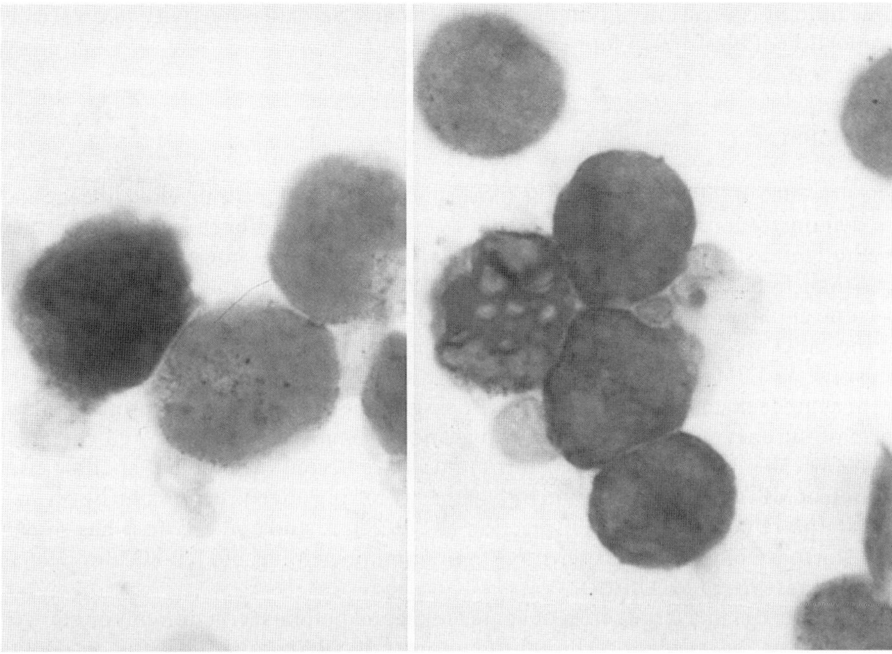

Fig. 11.2. Immunocytochemistry on a case of BAL with a T-lymphoid and myeloid phenotype. *Right* cells strongly positive with anti-MPO; *left* a blast cell positive with the T-cell marker CD3

were negative with specific B-lymphoid (CD79a, CD19, CD22) and T-lymphoid (CD3, CD2) markers. Cells (5–97%; median 40%) from 53 out of the 57 BAL cases (93%) were anti-MPO-positive (Table 11.2). These comprised 18 of 20 with a lymphoid or "ALL" morphology and cytochemistry MPO-negative and 30 out of 31 with an AML cytochemistry. The negative "AML" case corresponded to an acute monoblastic leukaemia AML-M5, also MPO-negative by light microscopy.

According to the immunophenotype, all the cases with a T and myeloid phenotype were anti-MPO-positive (Fig. 11.2) and 34 of the 36 with a B lymphoid and myeloid phenotype. The two cases with trilineage differentiation were also anti-MPO-positive whereas the remaining two with a dual B and T lymphoid phenotype were negative with the anti-MPO mAb. Expression of other myeloid markers (CD13, CD33, CD14) ranged from 89% to 20% of cases (Table 11.2). All cases expressed B or T lymphoid markers and the majority were positive with anti-HLA-Dr, CD34 and TdT.

11.3.2 Transmission Electron Microscopy (TEM) Cytochemistry

MPO activity was detected in the nuclear envelope, endoplasmic reticulum, Golgi apparatus and/or small cytoplasmic granules in all the AML-Mo cases investigated and in the 14 BAL cases. The latter comprised six cases classified as ALL, i.e. MPO-negative by light microscopy. In addition, the immunogold method demonstrated a concomitant expression of lymphoid antigens and peroxidase activity in a variable proportion of blasts from four cases.

11.4 Discussion

The presence of MPO represents the most specific marker for the myeloid lineage and its demonstration by light microscopic cytochemistry had been in the past a key method for distinguishing ALL from AML [1]. TEM has been shown to be more sensitive than light microscopy for detecting MPO [3, 6]. Over the last decade it has become apparent that mAbs that detect the active enzyme and the proenzyme form of the MPO are more sensitive than light microscopy cytochemistry and almost as sensitive as TEM for detecting the enzyme activity in poorly differentiated blasts from acute leukaemias. This is not surprising as the mAb detects the properoxidase form of the enzyme which is still inactive and hence not detectable by cytochemical reactions. In addition, molecular techniques have become available that allow estimation of MPO mRNA expression by Northern blotting or gene expression by reverse transcriptase polymerase chain reaction (RT-PCR). A study on cell lines has shown that Northern blotting is less sensitive than immunostaining and RT-PCR for detecting the presence of MPO [10].

Our findings in a large series of acute leukaemia patients with difficult or equivocal diagnosis by standard techniques demonstrate and confirm that the anti-MPO mAb is a highly sensitive means for detecting the myeloid commitment in blasts from poorly undifferentiated leukaemias. Thus, cells from close to half of AML-M0 cases, which are by definition MPO-negative by light microscopy, were reactive with anti-MPO. This was further confirmed by TEM that revealed the enzyme activity in a variable proportion of such cells. In addition, we have shown that this marker is essential to define the myeloid component in BAL cases, in which the immunophenotype reveals lymphoid and myeloid features and thus, are difficult to classify. In particular, MPO expression is important and relevant in those BAL cases with an ALL morphology to distinguish them from ALL with an aberrant expression of a myeloid marker, e.g. CD13. To this end, in the scoring system to define BAL, the anti-MPO mAb has the higher score for the myeloid lineage (2 points) according to the EGIL classification.

In addition to its sensitivity, we confirmed the specificity of the anti-MPO mAb as all cases of bona fide ALL were anti-MPO-negative and there was a good correlation between the TEM findings and expression of this marker.

11.5 Summary

The expression of MPO is the hallmark of the myeloid lineage. We describe the value of the anti-MPO mAb that detects the alpha-chain of the peroxidase and the proenzyme form in the diagnosis of acute leukaemias which are difficult to classify by standard methods. The reactivity with anti-MPO was assessed by immunocytochemistry using the APAAP method and/or flow cytometry after fixation and cell permeabilization. The 74 cases studied comprised 17 poorly differentiated AML (AML-M0), negative by light microscopy with the cytochemical reactions Sudan Black B (SBB), MPO and non-specific esterase (ANAE), and 57 cases with BAL with positive ("AML"-type) or negative ("ALL"-type) cytochemistry with blasts coexpressing lymphoid and myeloid antigens. The diagnosis of AML-Mo was made according to the FAB criteria and diagnosis of BAL according to the EGIL criteria. All BAL cases scored over 2 points and had the following phenotypes: B and myeloid phenotype (36/57), T and myeloid phenotype (17/57), B and T lymphoid phenotype (2/57) and trilineage differentiation (myeloid, B and T lymphoid) (2/57).

By morphology and standard light microscopic cytochemistry the BAL cases were classified as: "ALL" (20 cases) or "AML" (31 cases) (slides not available for review in 6 cases). Controls for the anti-MPO included cases of AML, ALL and normal bone marrow myeloid precursor cells. The anti-MPO mAb was expressed in most normal myeloid precursors and in 92% of cases of AML but was consistently negative in all bona fide ALL cases. In 8/17 AML-Mo cases (47%), anti-MPO was positive. All these cases were positive with one or two myeloid markers, either CD13 (13/17) and/or CD33 (10/17) and the majority expressed early haemopoietic stem-cell markers (CD34, HLA-Dr) whereas one-third and one-half of them were, respectively, TdT-positive, CD7-positive. Cells from most BAL cases (53/57, 93%) were anti-MPO-positive and these comprised 18 with a "lymphoblastic" morphology and cytochemistry and 30 with a "myeloid" morphology.

According to the immunophenotypes, the anti-MPO-positive BAL cases included 34/36 with a B-lymphoid and myeloid phenotype, 17 with a T and myeloid phenotype and the 2 with trilineage differentiation. Ultrastructural cytochemistry confirmed the presence of MPO activity in all AML-MO cases investigated and in the 14 BAL cases studied. In the latter, immunoelectron microscopy (immunogold method) showed in four patients a variable proportion of cells displaying lymphoid markers and MPO activity. These findings demonstrate that anti-MPO mAb is a sensitive and specific key marker for demonstrating myeloid commitment of undifferentiated blasts in acute leukaemias.

References

1. Bennett JM, Catovsky D, Daniel M-T, Flandrin G, Galton DAG, Gralnick HR, Sultan C (1976) Proposals for the classification of acute leukaemias. Br J Haematol 33: 451–458
2. Matutes E, Pombo de Oliveira M, Foroni L, Morilla R, Catovsky D (1988) The role of ultrastructural cytochemistry and monoclonal antibodies in clarifying the nature of undifferentiated cells in acute leukaemia. Br J Haematol 69: 205–211
3. Buccheri V, Shetty V, Yoshida N, Morilla R, Matutes E, Catovsky D (1992) The role of an anti-myeloperoxidase antibody in the diagnosis and classification of acute leukaemia: a comparison with light and electron microscopy cytochemistry. Br J Haematol 8: 62–68
4. Venditti A, Poeta G del, Buccisano F, Tamburini A, Cox MC, Stasi R, Bruno A, Aronica G, Maffei L, Suppo G, Simone MD, Forte L, Cordero V, Postorino M, Tufilli V, Isacchi G, Masi M, Papa G, Amadori R (1997) Minimally differentiated acute myeloid leukemia (AML-Mo): comparison of 25 cases with other French-American-British subtypes. Blood 89: 621–629
5. Campana D, Hansen-Hagge TE, Matutes E, Coustan-Smith E, Yokota R, Shetty V, Bartram CR, Janossy G (1990) Phenotypic, genotypic, cytochemical and ultrastructural characterization of acute undifferentiated leukemia. Leukemia 4: 620–624
6. Matutes E, Buccheri V, Morilla R, Shetty V, Dyer M, Catovsky D (1993) Phenotypic heterogeneity in acute leukemias: immunological characterization and clinical relevance. In: Ludwig W, Thiel E (eds) Recent advances in cell biology of acute leukemia. Impact on clinical diagnosis and therapy. Springer-Verlag, Berlin Heidelberg New York, pp 41–52

7. Bennett JM, Catovsky D, Daniel M-T, Flandrin G, Galton DAG, Gralnick HR, Sultan C (1991) Proposal for the recognition of minimally differentiated acute myeloid leukaemia. Br J Haematol 78: 325–329
8. Matutes E, Morilla R, Farahat N, Carbonell F, Swansbury J, Dyer M, Catovsky D (1997) Definition of acute biphenotypic leukemia. Haematologica 82: 64–66
9. Bene MC, Castoldi G, Knapp W, Ludwig W, Matutes E, Orfao A, Veer M van't, for The European Group for the Immunological Classification of Leukaemias (EGIL) (1995) Proposals for the immunological classification of acute leukemias. Leukemia 9: 1783–1786
10. Ma W, Hu ZB, Drexler HG (1994) Sensitivity of different methods for the detection of myeloperoxidase in leukemia cells. Leukemia 8: 336–342

Chapter 12

The Role of Myeloperoxidase in Myeloid Leukemia and Multiple Sclerosis

Wanda F. Reynolds

12.1 Introduction

The myeloperoxidase (MPO) gene is expressed specifically in the myeloid lineage [1–3] and in myeloid leukemias [4], which represent myeloid precursors blocked in their ability to differentiate. In developing myeloid cells in the bone marrow, MPO gene expression is highest in late myeloblasts and promyelocytes [1]. As these precursor cells mature along the granulocyte or monocyte lineages, MPO transcription decreases [3], while the MPO protein remains stored at high levels in neutrophils and monocytes in cytoplasmic lysosomes termed azurophilic granules. When blood monocytes or HL60 myeloid leukemic cells are induced in culture to differentiate into macrophages, MPO protein and mRNA levels are low to undetectable [2, 3, 5].

The MPO gene encodes an 83-kDa precursor polypeptide which is post-translationally processed to give rise to 13.5 kDa and 59 kDa subunits [3, 6]. Two of each subunit are assembled with heme groups to produce the tetrameric enzyme. MPO catalyzes the reaction of chloride and hydrogen peroxide to yield hypochlorous acid (HOCl), a strong oxidant [7]. In the presence of superoxide ($O_2^{\cdot -}$), HOCl generates highly reactive hydroxyl radicals. MPO also generates free chlorine (Cl_2) which acts as a potent halogenating agent [8]. MPO and its reactive byproducts have been linked to DNA strand breakage [9], generation of carcinogens [10, 11], and inhibition of DNA repair [12]. MPO has been linked to disease states mediated by later stage myeloid cells, neutrophils or monocytes/macrophages, including multiple sclerosis (MS) [13], myeloid leukemia [14, 15], atherosclerosis [16], and cystic fibrosis [17], indicating that the antimicrobial activity of MPO is accompanied by inadvertent cytotoxic effects.

Reactive oxygen species generated by macrophages have been implicated in many inflammatory and neurodegenerative disorders. Much of the research has focused on monocyte-macrophage-generated superoxide ion and nitric oxide. The potential contribution of the MPO enzyme has not been investigated, possibly due to a general conception that macrophages lack MPO [3]. However, findings summarized here suggest that MPO expression can be induced in macrophages in response to certain inflammatory stimuli.

12.2 The MPO Promoter Polymorphism

Two MPO alleles, termed Sp and N, differ at one known position (–463 G/A) in an upstream Alu sequence, within a cluster of hormone response elements (AluHRE) [14, 18, 19] (Fig. 12.1). This AluHRE consists of four hexamer halfsites, related to the

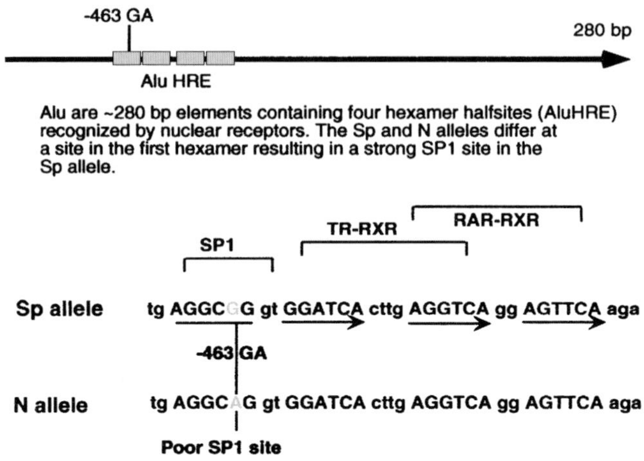

Fig. 12.1. Diagram indicates the position within the 280-bp Alu sequence of four hexamer halfsites recognized by retinoic acid receptors (RAR-RXR heterodimers), thyroid hormone receptors (TR-RXR), and SP1. The –463 G/A base difference distinguishing the Sp and N alleles is within the first hexamer and creates an SP1 site in the Sp allele, which is the site and hormone response elements in the Alu sequence preceding the Sp and N alleles

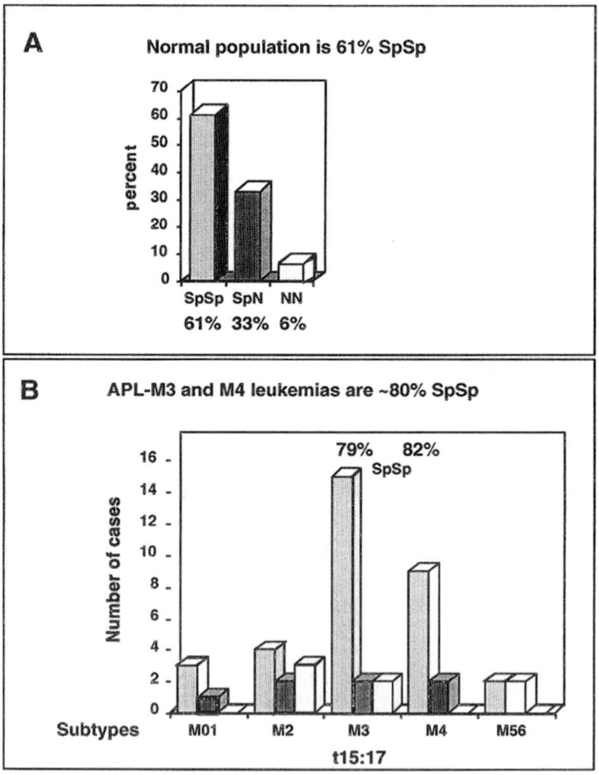

Fig. 12.2A,B. Diagram of the relative numbers of the MPO genotypes in the normal population (**A**) and myeloid leukemia patients (**B**)

consensus nuclear receptor binding site AGGTCA, oriented as direct repeats spaced by 2, 4, and 2 bp (DR-2–4-2). The 3' hexamer pair bind retinoic acid receptors and functions as a retinoic acid response element (RARE) in transfection assays, while the central DR4 pair functions as a thyroid hormone response element (TRE). The first DR-2 element contains the –463 G/A base difference. The –463 G in the Sp allele creates the core binding site for the SP1 transcription factor (GGCGGG), whereas the –463 A in this core site diminishes binding in the N allele (GGCAGG). This element from the Sp allele increases transcription of a reporter gene by 25-fold in cellular transfection assays, whereas the corresponding N allele element activates by only a fewfold [14]. Interestingly, the first hexamer pair in the Sp allele also functions as a strong RARE, possibly reflecting the recent finding that SP1 can tether the retinoic acid receptor (RAR) to a nearby halfsite.

The Sp allele is predominant in the general population, which is 61% SpSp, 33% SpN, and 6% NN [14, 15] (Fig. 12.2). As detailed below, the SpSp genotype is even more prevalent (80%) in certain myeloid leukemias [14, 15] and in MS. Moreover, the SpSp genotype is associated with higher MPO gene expression [15]. The overrepresentation of the higher expressing SpSp genotype in acute myeloid leukemia (AML) and MS implies that MPO is a causative agent in these disease states.

12.3 Over-representation of the SpSp Genotype in APL-M3 and M4 Myeloid Leukemias

AMLs are classified by the French-American-British (FAB) morphological scheme as subtypes M0 through M6, approximating early to later stages of myeloid development, with M0 and M1 characterized as myeloblastic, M3 as promyelocytic, and M5 as monocytic leukemia. The highest level of MPO expression is seen in acute promyelocytic leukemia (APL-M3) and can be up to 20 times higher than in Ml or MS subtypes [4]. This indicates that myeloid-specific transcription factors essential for MPO transcription are present at highest levels at promyelocyte stage. Because the AluHRE with the SP1 and RARE is a non-cell-type-specific enhancer, this element is unlikely to affect transcription in the absence of stage-specific factors required for MPO transcription. This predicts that the SpSp genotype, if linked to AML, might be more linked to subtypes around the promyelocyte stage which have the essential transcription factor background.

APL-M3 is characterized by the chromosomal translocation t [15, 17] which interrupts the RARα gene, fusing it to the PML gene which encodes a putative zinc finger transcription factor [20]. The PML-RAR fusion protein retains the DNA-binding and ligand-binding domains of RARα and most of the PML protein, and is thought to be a dominant-negative oncoprotein which interferes with the expression of genes involved in normal myeloid differentiation [20]. Disruption of the normal retinoic acid response appears to be key to the block in APL cell maturation. Treatment with all-*trans*-retinoic acid (RA) induces APL cells to proceed to differentiate into mature granulocytes, and accordingly, treatment of APL patients with RA results in complete remission in most cases. One event induced by RA treatment of APL cells is rapid shutdown of MPO transcription [21]. This raises the question as to whether the Alu-encoded RARE might be involved in the RA-mediated regulation of MPO expression.

We analyzed the MPO genotype in 46 AML cases classified as to subtypes to determine if the SpSp genotype is most associated with subtypes which express significant levels of MPO [15]. The APL-M3 cases were 79% SpSp (15 of 19 cases), and the M4 cases were 82% SpSp (9 of 11 cases). The eight cases categorized as low-expressing subtypes, M0,1,5,6, were 62% SpSp, equivalent to the normal population.

12.4 The SpSp Genotype Is Associated with Higher Levels of MPO mRNA Expression

The over-representation of the SpSp genotype in APL-M3 and M4 cases implies that this genotype has an increased risk of developing these leukemias. One possible explanation is that the SpSp genotype is linked to higher levels of MPO gene expression, resulting in increased DNA damage, thereby increasing the risk of leukemia. MPO comprises up to 5% of the cell mass of promyelocytes, so it is possible that the levels of MPO-generated radicals is sufficient to cause significant DNA damage. To directly determine if the SpSp genotype is linked to higher MPO gene expression, we compared the steady-state levels of MPO mRNA in primary AML cells of known MPO genotype. RNA was isolated by Trizol reagent and cDNA was prepared (First Strand Kit, Pharmacia) and used in PCR reactions with primers designed to amplify a region between position +30 and +580 of the MPO coding region. The different samples were normalized by comparison of beta-actin levels in separate RT-PCR reactions. Western analysis was carried out to compare MPO protein levels. Overall, the findings indicated that leukemic cells of the SpSp genotype contain two- to threefold higher levels of MPO mRNA and protein than do SpN or NN cells.

12.5 Evidence for MPO Involvement in Multiple Sclerosis

MS is an inflammatory disease of the central nervous system, characterized by multifocal lesions corresponding to sites of demyelination of axons. This autoimmune condition is initiated by T cells recognizing myelin associated antigens, such as myelin basic protein [22]. The cause in unknown but it is hypothesized that an infection by microbes antigenically related to myelin-associated proteins results in the appearance of cross-reactive T cells. Myelin-reactive T cells at lesions secrete cytokines/chemokines that induce infiltration and activation of macrophages and resident microglia. These phagocytes attack and degrade the myelin sheath, a process which may be exacerbated by their release of compounds which generate reactive oxygen species (ROS) [7, 23].

MS is estimated to strike as many as 1 in 2000 people. Two-thirds of cases are female, suggesting involvement of sex hormones. Although there is a clear genetic influence, as seen by the 30% concordance in identical twins [24], the only consistently observed feature detected by genetic linkage studies is the major histocompatibility complex. Disease states such as MS are likely to involve inherited arrays of genetic polymorphisms, such as the MPO polymorphism, each exerting modest or slight effect on disease progression [25].

To look for evidence of differential association of the MPO alleles with MS cases, we analyzed the genotypes of 59 MS cases of which 43 (73%) were female, consistent

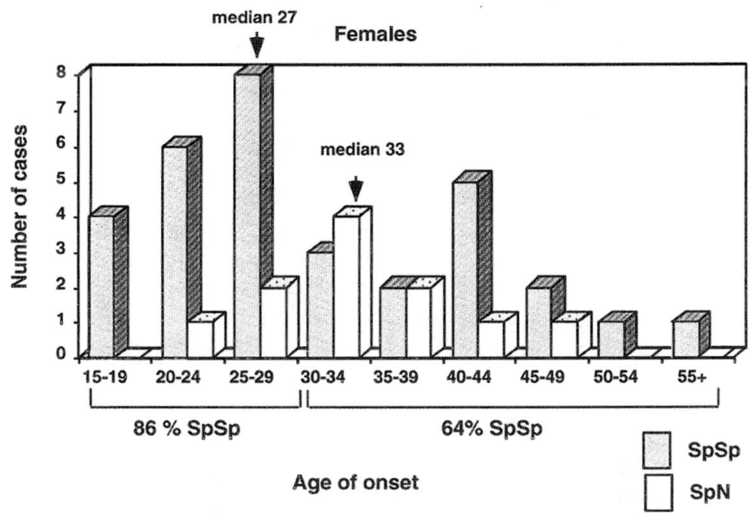

Fig. 12.3. Over-representation of the SpSp MPO genotype in early-onset MS in females. The histogram indicates the number of female MS patients who had first onset of disease symptoms at the ages indicated. The *light shaded columns* represent 11 SpN-NN female patients, while the *dark columns* represent 32 SpSp female patients. The median age of onset for each group is indicated by the *arrows*. *Brackets* at the bottom indicate that patients who had onset prior to age 30 years were 86% SpSp, while those with onset at or over age 30 years were 64% SpSp

with the known prevalence of females in MS. The percentage SpSp genotype among the female MS cases was 74%, higher than the 61% SpSp observed in the general population. This over-representation was more pronounced in females with early onset of disease symptoms, prior to age 30 years, which were 86% SpSp (Fig. 12.3). The median onset age for SpSp cases was 27 years, as compared to 33 years for the SpN cases. Thus the SpSp genotype is associated with increased incidence and earlier onset of MS symptoms in females.

12.6 Immunocytochemical Detection of MPO Protein in Microglia/Macrophages in MS Brain Lesions

Because MPO is myeloid-specific, and macrophages are known to play a key role in the demyelination, macrophages are the likely source of MPO at MS lesions. To determine if MPO-positive macrophages are present at MS lesions, immunocytochemical analysis was performed on sectioned brain tissue using polyclonal antibodies to human MPO, and antibodies to macrophage-specific antigens, HLA-DR and CD68. Lesions were identified as areas devoid of myelin basic protein, as indicated by antibody staining, or with oil-red-O, which stains phagocytic vesicles filled with partially degraded myelin within macrophages. Significant numbers of MPO-positive macrophage-microglia were found in lesions, in some areas containing large oil-red-O-stained vesicles. A total of 32 lesions were examined from five different donors. MPO-positive cells were detected in approximately 70% of the active lesions in all five donors. MPO-positive cells were not observed in sections from normal brain tissue outside of the circulatory system.

12.7
Detection of MPO mRNA Sequences in cDNA from Human Microglia

As further evidence that brain macrophages have the capacity to express MPO, MPO mRNA sequences were detected by RT-PCR in primary human microglia isolated from surgically excised brain tissue. More recent studies in our laboratory have detected MPO mRNA sequences in a rodent microglia cell line and in thioglycolate-activated peritoneal macrophages from mice.

One model concerning MPO involvement in MS is that myelin-reactive T cells emit cytokines that attract and activate microglial macrophages in lesions, reactivating or upregulating MPO gene expression. MPO-generated HOCl creates reactive oxidizing radicals that contribute to the destruction of the oligodendrocytic myelin sheath surrounding axons. Individuals with two copies of the Sp allele express more MPO protein, and thereby sustain more damage at lesions, leading to an accelerated disease course. The MPO-positive cells detected in MS brain lesions could be resident microglia or invading blood-borne macrophages, since there is no method to distinguish unequivocally between these two macrophage types. Both are detected in or around MS lesions and both are implicated in MS pathology [5]. The detection of MPO mRNA sequences in isolated microglia argues that this macrophage type has the ability to express the gene, although blood-derived macrophages might also. These two macrophage types may be essentially interchangeable. Upregulation of MPO gene expression in activated macrophages would seem consistent with the presumed antimicrobial function of this enzyme pathway. Considering the cytotoxic nature of MPO byproducts, the storage of this enzyme in lysosomes in quiescent microglia throughout the brain parenchyma might represent a serious risk to nondividing neurons, such that this gene is only turned on in response to cytokines signaling an active infection or other serious insult to the brain. For blood-borne monocytes and neutrophils, the inadvertent release of lysosomal stores of MPO may represent less of a cytotoxic risk, while allowing more immediate response to infection. For these reasons, MPO may be differentially expressed or regulated in brain microglia as compared to other macrophage types.

The association of MPO with MS pathology as well as myeloid leukemia raises the possibility that this enzyme plays a role in other autoimmune or inflammatory disorders in which microglia or macrophages are implicated, such as Alzheimer's disease (AD), diabetes, asthma, or psoriasis. Potential therapeutics could be aimed specifically at the MPO enzyme based on the observation that there are genetically MPO-null or deficient individuals who have no apparent MPO-related health defects [26].

12.8
Colocalization of MPO with Amyloid Beta in Alzheimer's Plaques

More recent findings implicate MPO in AD [27]. One characteristic of Alzheimer's plaques is a surrounding cluster of microglia which are attracted to and activated by amyloid beta, a 42 amino acid proteolysis product which is the primary component of amyloid plaques. It is suspected that the activated microglia release oxidizing species which promote amyloid deposition or otherwise contribute directly to nearby neuron damage. Antibodies have been used to detect MPO in CD68-positive microglia encircling the plaques in AD brain sections, and confocal images demonstrate that MPO colocalizes with beta amyloid in such plaques. Moreover, synthetic beta amyloid peptides have been found to induce MPO gene expression in the BV2 mouse microglia cell line. These findings suggest that MPO gene expression is induced in microglia in proximity to plaques. As evidence that MPO levels influence the incidence of AD, the SpSp genotype is over-represented in females with AD, as previously observed in females with MS. There is also an interesting correlation between the amount of MPO deposition and the ApoE genotype. ApoE is a lipid transport molecule represented by several alleles, E1–E4. The ApoE4 allele is associated with increased AD risk, as well as increased amyloid deposition. Patients who carry one or two ApoE4 alleles have been found to have the highest amounts of MPO deposition at plaques. This suggests a functional synergy between MPO and ApoE related to amyloid deposition.

12.9 Summary

Two MPO alleles, Sp and N, differ at –463 G/A in the upstream promoter, creating an SP1 site in the Sp allele, lacking in the N allele. The SP1 site overlaps hormone response elements within an Alu element (AluHRE). Transfection assays show the Sp allele AluHRE is the stronger transcriptional activator. Moreover, primary myeloid leukemia cells of the SpSp genotype express higher levels of MPO mRNA and protein than SpN or NN cells. The SpSp genotype is predominant in the normal population (61%) and is over-represented (80%) in APL (t15;17) and AML-M4. This over-representation suggests that higher MPO levels increases the risk of developing leukemia, possibly due to MPO-generated oxidizing radicals which damage DNA. Similarly, in MS, SpSp females are over-represented (74%), especially in early-onset disease (86%), suggesting that increased MPO expression in brain macrophages accelerates damage to the myelin sheath. Immunohistochemical staining reveals MPO-positive macrophages at active MS lesions but not in normal brain. Also, MPO mRNA can be detected in cDNA from human and murine primary microglia. Most recent evidence indicates that MPO is present in Alzheimer's plaques and in activated microglia surrounding plaques. Moreover, the SpSp genotype is over-represented in females with AD, as in MS. These findings provide evidence that MPO plays a significant role in disease states mediated by myeloid cells.

Acknowledgements. This study was funded by a grant from the National Cancer Institute at NIH (ROl CA72995).

References

1. Fouret P, du Bois RM, Bemaudin JF, Takahashi H, Ferrans VJ, Crystal RG (1989) Expression of the neutrophil elastase gene during human bone marrow cell differentiation. J Exp Med 169: 833–845
2. Koeffler HP, Ranyard J, Pertcheck M (1985) Myeloperoxidase: its structure and expression during myeloid differentiation. Blood 65: 484–491
3. Tobler A, Miller CW, Johnson KR, Selsted ME, Rovera G, Koeffler HP (1988) Regulation of gene expression of myeloperoxidase during myeloid differentiation. J Cell Physiol 136: 215–225
4. Zaki SR, Austin GE, Swan D, Srinivasan A, Ragab AH, Chan WC (1989) Human myeloperoxidase gene expression in acute leukemia. Blood 74: 2096–2102
5. Ulvestad E, Williams K, Mork S, Antel J, Nyland H (1994) Phenotypic differences between human monocytes/macrophages and microglial cells studied in situ and in vitro. J Neuropathol Exp Neurol 53: 492–505
6. Nauseef WM, Cogley M, McCormick S (1996) Effect of the R569W missense mutation of the biosynthesis of myeloperoxidase. J Biol Chem 271: 9546–9549
7. Klebanoff SJ (1980) Oxygen metabolism and the toxic properties of phagocytes. Ann Intern Med 93: 480–489
8. Hazen S, Hsu F, Duffin K, Heinecke JW (1996) Molecular chlorine generated by the myeloperoxidase-hydrogen peroxide-chloride system of phagocytes converts low density lipoprotein cholesterol into a family of chlorinated sterols. J Biol Chem 271: 23080–23088
9. Van Rensburg CE, Van Staden AM, Anderson R, Van Rensburg EJ (1992) Hypochlorous acid potentiates hydrogen peroxide-mediated DNA-strand breaks in human mononuclear leucocytes. Mutat Res 265: 255–261
10. Petruska JM, Mosebrook DR, Jakab GJ, Trush MA (1992) Myeloperoxidase enhanced formation of trans-7,8-dihydroxibenzo(a)pyrene-DNA adducts in lung tissue in vitro: a role of pulmonary inflammation in the bioactivation of a procarcinogen. Carcinogenesis 13: 1075–1081
11. Josephy PD (1996) The role of peroxidase-catalyzed activation of aromatic amines in breast cancer. Mutagenesis 11: 3–7
12. Pero RW, Sheng Y, Olsson A, Bryngelsson C, Lund-Pero M (1996) Hypochlorous acid/N-chloramines are naturally produced DNA repair inhibitors. Carcinogenesis 17: 13–18
13. Nagra R, Becher B, Tourtellotte WW, Antel JJ, Gold D, Smith RA, James R. Nelson JR, Reynolds WF (1997) Immunohistochemical and genetic evidence of myeloperoxidase involvement in multiple sclerosis. J Neuroimmunol 78: 97–107
14. Piedrafita FJ, Molander R, Vansant G, Orlova EA, Pfahl M, Reynolds WF (1996) An Alu element in the myeloperoxidase promoter contains a composite SP1-thyroid hormone-retinoic acid response element. J Biol Chem 271: 14412–14420
15. Reynolds WF, Chang E, Douer D, Ball ED, Kanda V (1997) An allelic association implicates myeloperoxidase in the etiology of acute promyelocytic leukemia. Blood 90: 2730–2737
16. Daugherty A, Rateri DL, Dunn J, Heinecke JW (1994) Myeloperoxidase: a catalyst for lipoprotein oxidation, is expressed in human atherosclerotic lesions. J Clin Invest 94: 437–444
17. Witko-Sarsat V, Allen R, Paulais M, Nguyen A, Bessou G, Lenoir G, Descamps-Latscha B (1996) Disturbed myeloperoxidase dependent activity of neutrophils in cystic fibrosis homozygotes and heterozygotes and its correction by amiloride. J Immunol 157: 2728–2735
18. Vansant G, Reynolds WF (1995) The consensus sequence of a major Alu subfamily contains a functional retinoic acid response element. Proc Natl Acad Sci U S A 18: 8229–8233
19. Austin GE, Lam L, Zaki SR, Chan WC, Hodge T, Hou J, Swan D, Zhang W, Racine M, Whitsett C, Brown T (1993) Sequence comparison of putative regulatory DNA of the 5' flanking region of the myeloperoxidase gene in normal and leukemic bone marrow cells. Leukemia 9: 1445–1450
20. The H de, Lavau C, Marchio A, Chomienne C, Degos L, Dejean A (1991) The PML-RAR alpha fusion mRNA generated by the t(15;17) translocation in acute promyelocytic leukemia encodes a functionally altered RAR. Cell 66: 675–681
21. Douer D, Preston-Martin S, Chang E, Nichols PW, Watkins KJ, Levine AM (1996) High frequency of acute promyelocytic leukemia among Latinos with acute myeloid leukemia. Blood 87: 308–312
22. Steinman L (1996) Multiple sclerosis: a coordinated immunological attack against myelin in the central nervous system. Cell 85: 299–302
23. Halliwell B (1992) Oxygen radicals as key mediators in neurological disease: fact or fiction? Ann Neurol 32s: 10–15
24. Ebers GC, Sadovnick AD, Risch NJ, and the Canadian Collaborative Study Group (1995) A genetic basis for familial aggregation in MS. Nature 377: 150–151
25. Risch N, Merikangas K (1996) The future of genetic studies of complex human diseases. Science 273: 1516–1517

26. Nauseef WM, Root RK, Malech HL (1983) Biochemical and immunologic analysis of hereditary myeloperoxidase deficiency. J Clin Invest 71: 1297–1307
27. Reynolds WF, Rhees J, Maciejewski D, Paladino T, Sieburg H, Maki RA, Masliah E (1999) Myeloperoxidase polymorphism is associated with gender specific risk for Alzheimer's disease. Exp Neurol 155: 31–41

Chapter 13

Pathways for Oxidative Tissue Injury by Myeloperoxidase

Jay W. Heinecke

13.1 Introduction

Oxidant generation by phagocytes may represent an important pathway for tissue damage in pathological conditions ranging from atherosclerosis to ischemia reperfusion injury and cancer. The pathway for oxidant generation by activated phagocytic white blood cells begins with a membrane-associated NADPH oxidase that generates superoxide [1, 2]. Superoxide spontaneously or enzymatically dismutates to form hydrogen peroxide, a relatively unreactive oxidant. However, the oxidative potential of hydrogen peroxide is amplified by myeloperoxidase, a secreted heme protein [1, 2]. The enzyme represents an attractive candidate for monitoring phagocyte-mediated damage because it generates several distinct oxidants that covalently modify cellular targets.

This review summarizes recent progress in our understanding of the oxidation chemistry of myeloperoxidase.

13.2 Enzymology of Myeloperoxidase

Myeloperoxidase is one of the most abundant proteins in phagocytes: it represents about 5% of neutrophil protein and about 1% of monocyte protein [3]. Recent studies suggest that the enzyme is also expressed by human macrophages in vivo [4]. Myeloperoxidase reacts with hydrogen peroxide to form "compound I", a π-cation radical complex that is reduced to the native state by the oxidation of halide and other substrates [2]. Regeneration of the native enzyme requires reduction by: (1) a single two-electron transfer reaction, or (2) two sequential one-electron transfer reactions. Because the active site of myeloperoxidase is buried in a protein cleft and inaccessible to large macromolecules, low molecular weight intermediates are required to convey oxidizing equivalents from its heme group to the target for damage.

The major product of myeloperoxidase at plasma concentrations of chloride ion (Cl^-) is hypochlorous acid (HOCl) [5]:

$$H_2O_2 + Cl^- + H^+ \rightarrow HOCl + H_2O \qquad (1)$$

This potent cytotoxin may be viewed as Cl^+, which is formally equivalent to the two-electron oxidation product of Cl^-. HOCl chlorinates electron-rich substrates and oxidatively bleaches heme proteins, nucleotides and carotenoids [1, 2, 6], events which may be important in bacterial killing by myeloperoxidase. Chlorinated compounds represent attractive candidates for monitoring phagocyte-mediated oxidative damage because myeloperoxidase is the only human enzyme known to generate HOCl at plasma concentrations of halide.

13.3 Dityrosine Synthesis

Another potential substrate for myeloperoxidase is the phenolic amino acid tyrosine, which readily undergoes one-electron oxidation to form tyrosyl radical [7]. The productive interaction of two tyrosyl radicals yields *o,o'*-dityrosine, an intensely fluorescent compound [8]. To explore the potential role of tyrosyl radical in the chemistry of myeloperoxidase, we studied the ability of the enzyme to synthesize dityrosine.

Isolated myeloperoxidase rapidly converts tyrosine to dityrosine by a reaction which requires hydrogen peroxide [9]. Phorbol ester-stimulated neutrophils and macrophages similarly generate dityrosine from tyrosine. This reaction is inhibited by catalase (a scavenger of hydrogen peroxide) and heme poisons, indicating that the cellular pathway is dependent upon myeloperoxidase. These results indicate that activated phagocytes employ the myeloperoxidase-hydrogen peroxide system to generate tyrosyl radical.

13.4 Protein Dityrosine Crosslinking

The production of tyrosyl radical by myeloperoxidase raised the possibility that proteins might be one target for damage. To explore this possibility we exposed albumin to the myeloperoxidase-hydrogen peroxide system, and then analyzed the albumin for protein-bound dityrosine [10]. In the absence of tyrosine there was little modification of albumin. Inclusion of tyrosine in the reaction mixture lead to a marked increase in protein-bound dityrosine-like fluorescence. To firmly establish the structure of the fluorescent oxidation product we subjected the material isolated by ion exchange chromatography to gas chromatography-mass spectrometry. Both the retention time and mass spectrum of the compound were virtually identical to those of authentic dityrosine, conclusively identifying the major fluorescent product in myeloperoxidase-modified albumin as dityrosine.

Synthesis of protein-bound dityrosine by myeloperoxidase requires active enzyme, hydrogen peroxide and tyrosine; it is inhibited by heme poisons and the hydrogen peroxide scavenger catalase. Activated neutrophils similarly modify albumin. As with myeloperoxidase, the reaction requires tyrosine and is inhibited by heme poisons and catalase, strongly implicating myeloperoxidase in the reaction.

Collectively, these results indicate that human neutrophils employ the myeloperoxidase-hydrogen peroxide system to oxidatively crosslink proteins by a reaction involving tyrosyl radical. The phenolic coupling reaction is independent of free metal ions but requires tyrosine, implying that tyrosyl radical is serving as a diffusible catalyst that conveys oxidizing equivalents from the heme group to protein tyrosyl residues [10, 11].

13.5 Reactive Aldehydes

When myeloperoxidase oxidizes chloride, dityrosine formation is suppressed [9]. Therefore, in plasma, where Cl^- concentrations are high, myeloperoxidase may convert tyrosine to products other than dityrosine. Indeed, indirect evidence has long suggested that amino acids are decarboxylated and deaminated to form compounds with reactive carbonyls [12].

To explore this possibility, we exposed tyrosine to myeloperoxidase, hydrogen peroxide and Cl^- and analyzed the reaction products by high-performance liquid chromatography [13]. In the presence of the complete system, a single major oxidation product appeared. Using mass spectrometry, Fourier transform infrared spectroscopy, and high resolution NMR spectroscopy, we identified the isolated product as a highly reactive aldehyde, *p*-hydroxyphenylacetaldehyde (*p*HA). At plasma concentrations of Cl^- and tyrosine, *p*HA production consumed more than 80% of the hydrogen peroxide in the reaction mixture.

This reaction also proceeded readily in the presence of activated neutrophils. With optimal stimulation, *p*HA production accounted for over half of the cellular output of hydrogen peroxide [13]. Reagent HOCl also converted tyrosine into *p*HA, strongly implicating HOCl in the cellular reaction pathway. Using isotope dilution gas chromatography/mass spectrometry, we detected a covalent adduct of *p*HA and lysine in inflammatory tissues [14]. This strongly suggests that myeloperoxidase generates reactive aldehydes in vivo.

Recent studies indicate that myeloperoxidase converts virtually all of the common amino acids to reactive aldehydes [15, 16]. For example, threonine is converted into the potent cytotoxin acrolein [15]. Reactive aldehydes are thought to be of central importance in the genesis of vascular disease [17]. Aldehydes derived from lipid peroxidation convert low-density lipoprotein (LDL) to a ligand for the macrophage scavenger receptor, which plays a key role in the formation of foam cells, the cellular hallmark of atherosclerosis. Oxidation of proteins and lipids by glucose (an aldehyde in its open-chain form) during chronic hyperglycemia may play a similar role in crosslinking arterial wall proteins with plasma components, an event that may accelerate the vascular disease of diabetes mellitus. Reactive aldehydes generated by myeloperoxidase may similarly damage biomolecules at sites of acute and chronic inflammation.

13.6 Molecular Chlorine

Stable chlorinated products should serve as specific markers of oxidation by myeloperoxidase. We have demonstrated that one such product is 3-chlorotyrosine, which is produced by myeloperoxidase in a reaction that requires active enzyme, peroxide and chloride ion. Surprisingly, tyrosine chlorination by myeloperoxidase is optimal at acidic pH [18]. The pH dependence is not due to the enzymatic activity of myeloperoxidase since the yield of 3-chlorotyrosine with reagent HOCl increases with increasing $[H^+]$. Because the pK_a for $HOCl/ClO^-$ is about 7.4, this observation suggests that HOCl is not the chlorinating intermediate. HOCl is also in equilibrium with Cl_2 via a reaction that requires H^+ and Cl^-:

$$HOCl + Cl^- + H^+ \Leftrightarrow Cl_2 + H_2O \quad (2)$$

To determine whether HOCl or Cl_2 is the chlorinating intermediate, we examined the reaction requirements for the chlorination of tyrosine by HOCl [18]. Generation of Cl_2 should require Cl^- (scheme 2). Indeed, reagent HOCl failed to chlorinate tyrosine in the absence of Cl^-. In addition, the reaction was optimal under acidic conditions, consistent with a requirement for H^+ in the reaction. Finally, at neutral pH and in the absence of Cl^-, molecular chlorine readily generates 3-chlorotyrosine. These results strongly suggest that Cl_2, and not HOCl, is the chlorinating intermediate in tyrosine oxidation by myeloperoxidase.

To test the physiological relevance of molecular chlorine generation by myeloperoxidase, we determined whether human neutrophils generate 3-chlorotyrosine [18]. Phagocytosis of tyrosine encapsulated in opsonized red blood cells results in 3-chlorotyrosine formation. In addition, human neutrophils that are activated by adherence to fibronectin-coated glass surfaces also convert tyrosine into 3-chlorotyrosine. These results suggest that a Cl_2-like oxidant is generated by activated human neutrophils during phagocytosis.

Collectively, these results indicate that the myeloperoxidase-hydrogen peroxide-Cl^- system generates a Cl_2-like species at acidic pH. A number of such acidic compartments may exist in vivo, including the space between the closely juxtaposed membranes of adherent phagocytes and endothelium, as well as hypoxic tissue such as atherosclerotic lesions.

13.7
Myeloperoxidase in Human Atherosclerosis

We have previously shown that myeloperoxidase is present in human atherosclerotic lesions, where it colocalizes in part with tissue macrophages [4]. Many lines of evidence indicate that LDL, the major carrier of blood cholesterol, must be oxidized to promote the pathological events to atherosclerosis [17]. To explore the role of one myeloperoxidase product, tyrosyl radical, in promoting LDL oxidation in vivo, we used isotope dilution gas chromatography/mass spectrometry to quantify the level of dityrosine in lesional LDL [19]. We detected a remarkable 100-fold increase in dityrosine levels in lesional LDL compared with those in circulating LDL. These results suggest that tyrosyl radical, perhaps generated in part by myeloperoxidase, is one agent for LDL oxidation in the human artery wall.

To obtain direct evidence that myeloperoxidase promotes oxidation reactions in vivo, we used isotope dilution gas chromatography/mass spectrometry to quantify levels of 3-chlorotyrosine in lesional LDL. Lesional LDL contained 30-times more 3-chlorotyrosine than circulating LDL. These results provide strong evidence that halogenation reactions catalyzed by myeloperoxidase constitute one pathway for protein oxidation in vivo [20]. Moreover, they suggest that myeloperoxidase may play a critical role in rendering LDL atherogenic.

13.8 Summary

Activated phagocytes have been implicated in tissue damage in diseases ranging from arthritis to inflammatory bowel disease to ischemia-reperfusion injury. Our results suggest that one pathway for oxidative damage in the artery wall involves myeloperoxidase. We speculate that reactive intermediates generated by myeloperoxidase play a role in the pathogenesis of many inflammatory disorders.

References

1. Klebanoff SJ (1980) Oxygen metabolism and the toxic properties of phagocytes. Ann Intern Med 93: 480–489
2. Hurst JK, Barrette WC Jr (1989) Leukocytic oxygen activation and microbicidal oxidative toxins. CRC Crit Rev Biochem Mol Biol 24: 271–328
3. Nauseef WM (1988) Myeloperoxidase deficiency – phagocytic defects I. Hematology/oncology. Clin North Am 2: 135–158
4. Daugherty A, Dunn JL, Rateri D, Heinecke JW (1994) Myeloperoxidase, a catalyst for lipoprotein oxidation, is expressed in human atherosclerotic lesions. J Clin Invest 94: 437–444
5. Harrison JE, Schultz J (1976) Studies on the chlorinating activity of myeloperoxidase. J Biol Chem 251: 1371–1374
6. Thomas EL, Grisham MB, Jefferson MM (1986) Preparation and characterization of chloramines. Methods Enzymol 132: 569–585
7. Prince RC (1988) Tyrosine radicals. Trends Biochem Sci 13: 286–288
8. Anderson SO (1966) Covalent cross-links in a structural protein, Resilin. Acta Physiol Scand 66: 1–81
9. Heinecke JW, Li W, Daehnke HL, Goldstein JA (1993) Dityrosine, a specific marker of oxidation, is synthesized by the myeloperoxidase-hydrogen peroxide system of human neutrophils and macrophages. J Biol Chem 268: 4069–4077
10. Heinecke JW, Li W, Francis GA, Goldstein JA (1993) Tyrosyl radical generated by myeloperoxidase catalyzes the oxidative crosslinking of proteins. J Clin Invest 91: 2866– 2872
11. Francis GA, Mendez AJ, Bierman EL, Heinecke JW (1993) Oxidative tyrosylation of high density lipoprotein by peroxidase enhances cholesterol removal from cultured fibroblasts and macrophage foam cells. Proc Natl Acad Sci U S A 90: 6631–6635
12. Babior BM (1978) Oxygen-dependent microbial killing by phagocytes. N Engl J Med 298: 659–663
13. Hazen SL, Hsu FF, Heinecke JW (1996) p-Hydroxyphenylacetaldehyde is the major product of tyrosine oxidation by activated human phagocytes. J Biol Chem 271: 1861–1867
14. Hazen SL, Hsu FF, d'Avignon A, Anderson MM, Heinecke JW (1998) Human neutrophils employ myeloperoxidase to convert alpha-amino acids into a battery of reactive aldehydes. Biochemistry 37: 6864–6873
15. Anderson MM, Hazen SL, Hsu FF, Heinecke JW (1997) Human neutrophils employ the myeloperoxidase-hydrogen peroxide-chloride system to convert hydroxy amino acids into glycoaldehyde, 2-hydroxypropanal and acrolein. J Clin Invest 99: 424–432
16. Hazen SL, Gaut JP, Hsu FF, Crowley JR, d'Avignon A, Heinecke JW (1997) pHydroxyphenylacetaldehyde, the major product of l-tyrosine oxidation by the myeloperoxidase-H_2O_2-chloride system of phagocytes, covalently modifies epsilon-amino groups of protein lysine residues. J Biol Chem 272: 16990–16998
17. Berliner JA, Heinecke JW (1996) The role of oxidized lipoproteins in atherogenesis. Free Radic Biol Med 20: 707–727
18. Hazen SL, Hsu FF, Mueller DM, Crowley JR, Heinecke JW (1996) Human neutrophils employ chlorine gas as an oxidant during phagocytosis. J Clin Invest 98: 1283–1289
19. Leeuwenburgh C, Rasmussen JE, Hsu FF, Mueller DM, Pennathur S, Heinecke JW (1997) Mass spectrometric quantification of markers for protein oxidation by tyrosyl radical, copper, and hydroxyl radical in low density lipoprotein isolated from human atherosclerotic plaques. J Biol Chem 272: 3520–3526
20. Hazen SL, Hsu FF, Mueller DM, Crowley JR, Heinecke JW (1997) 3-Chlorotyrosine, a specific marker of myeloperoxidase-catalyzed oxidation, is markedly elevated in low density lipoprotein isolated from human atherosclerotic intima. J Clin Invest 99: 2075–2081

Chapter 14

Myeloperoxidase Activity of Neutrophils in Cystic Fibrosis

Véronique Witko-Sarsat

14.1 Introduction

Cystic fibrosis (CF) is an autosomal recessive disorder, common in Caucasians, which is characterized by increased sodium and chloride concentrations in sweat and by pancreatic insufficiency. CF patients eventually develop a chronic condition, characterized by production of thick sputum, impaired mucociliary clearance, and ultimately multiple bronchopulmonary infections [1, 2]. Thus the airway inflammation typical of CF reflects both an exacerbation of neutrophil-mediated tissue damage and a concomitant persistence of inflammation. In this context, our work has focused on the mechanisms underlying inflammation in CF and the critical assessment of the role of myeloperoxidase (MPO), a myeloid protein with both antimicrobial and proinflammatory activities.

14.2 Pathophysiology and Genetic Basis of CF

CF is caused by mutation of the cystic fibrosis transmembrane regulator (CFTR) gene. Over 100 mutations of CFTR are associated with CF, but the most common is the deletion of phenylalanine at position 508 of the predicted 1480-residue CFTR protein. CFTR is currently thought to act as a chloride channel on the apical surface of epithelial cells, and to respond to increases in intracellular levels of cyclic AMP [3, 4]. One characteristic feature in CF pathophysiology is airway inflammation with a predominant neutrophil infiltration, and early bacterial infection with *Staphylococcus aureus* and chronic colonization of *Pseudomonas aeruginosa* (Fig. 14.1). These infections generally provoke the activation of cells of the specific and nonspecific immune system and the migration of blood neutrophils to the lung. The chronic pulmonary inflammation and recurrent infection culminate in compromised lung function and premature death of CF patients.

14.3 Aim of Our Study

The mechanisms leading to neutrophil-dominated inflammation in CF patients as well as the relationships with CFTR mutations remain largely unknown. Prospects for effective gene therapy of the respiratory effects of CF are improving. Nonetheless, for the clinical management of CF it is extremely important both to investigate the causes of neutrophil recruitment, and to manage chronic active pulmonary inflammation.

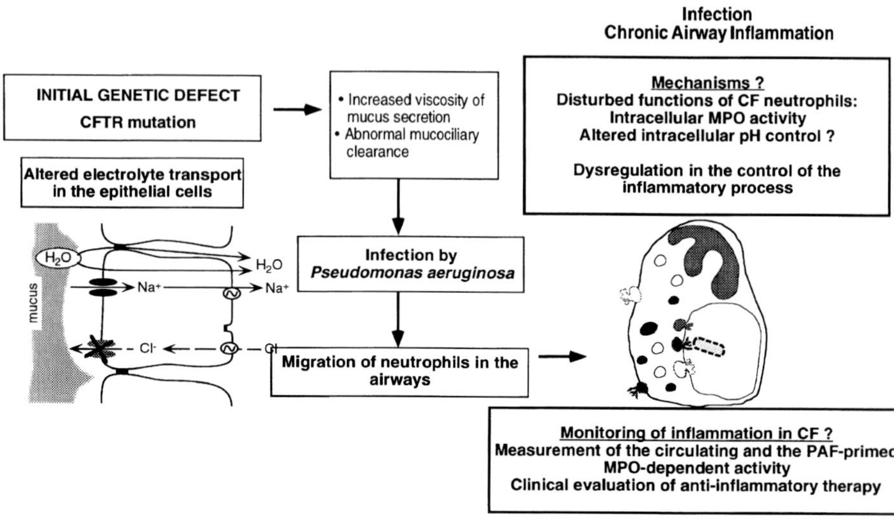

Fig. 14.1. Role of neutrophils in the airway inflammation associated with cystic fibrosis: a revisited picture. The classical pathophysiological mechanisms of neutrophil-dominated airway inflammation begin with the initial genetic defect which is the mutation of CFTR at the epithelial level thus modifying the ionic exchanges and impairing the mucociliary clearance. The consequence is an increased susceptibility to infections and especially with *P. aeruginosa*, leading to neutrophil recruitment and establishment of chronic airway inflammation. However, this classical picture does not take into account the recent finding of the presence of neutrophils in the airways in the absence of infection. Thus, according to recent research on CFTR functions and on CF neutrophils, it is tempting to propose that CF neutrophils could be genetically modified and show an exacerbated inflammatory response as illustrated by a constitutive increase in MPO activity in CF homozygotes and heterozygotes. This exacerbated inflammatory responses could proceed via other mechanisms which remain to be elucidated

Indeed, recent clinical studies have demonstrated a beneficial effect of nonsteroidal antiinflammatory drugs such as ibuprofen on the progression of the lung disease [5]. Our laboratory has focused especially on the role of MPO in the pathogenesis of CF in order to address several questions: (1) Is there a disturbance in MPO-derived oxidant formation in neutrophils from CF patients and is there a possible linkage with the CFTR mutation? (2) Could chlorinated oxidants play a role at the inflammation site? (3) What is the influence of bacterial infections on MPO activity in CF patients and can MPO activity be used to monitor inflammation?

14.4 Neutrophil-Dominated Airway Inflammation in Cystic Fibrosis

Influx of neutrophils into the airways and chronic respiratory bacterial infection are two prominent pulmonary features of CF. Interestingly, the neutrophilic inflammatory response precedes any microbiological evidence of bacteria in the airway [6]. Although neutrophils are essential for host defense, their unchecked recruitment and activation can progressively damage airways by virtue of the potent proinflammatory mediators that they generate. Most clinical studies designed to unravel the mechanisms causing CF patients to be highly susceptible to infections have focused on proteases and on serum-mediated bacteria opsonizing dysfunction. The current explanation for the persistence of *P. aeruginosa* in the respiratory tract is the emergence of mucoid strains which are resistant to phagocytosis and are specifically associated with CF [7]. Involved airways contain high concentrations of neutrophil-derived mediators, including inflammatory cytokines such as IL-8 [8] and TNF-α, metalloproteases such as gelatinase, serine proteases including elastase, cathepsin G [9] and proteinase 3 [10], and antibiotic peptides, defensins. It has been suggested that the antibiotic activity might be decreased in CF lung because of the modification in the ionic concentration in bronchial secretions [11]. Little attention has been given to the role of neutrophil-derived oxidants in CF. An overall oxidant/antioxidant imbalance has been observed in CF patients, based on evidence from plasma antioxidant status and total radical trapping antioxidant potential [12]. Although neutrophil-derived oxidants are probably involved in the onset of pulmonary inflammation in CF, this process is poorly understood.

4.5 MPO Activity in Neutrophils from CF Homozygotes and CF Heterozygotes

Activation of the neutrophil NADPH oxidase generates superoxide anion (O_2^-) which can dismutate into hydrogen peroxide (H_2O_2). MPO, an enzyme contained in azurophil granules, catalyzes the H_2O_2-dependent oxidation of chloride (Cl^-), which yields singlet oxygen ($1O_2$) and hypochlorous acid (HOCl). The latter can react with amines to yield chloramines, the so-called long-lived oxidants. We analyzed oxidative metabolism in neutrophils from uninfected CF children, their respective parents (asymptomatic heterozygote CF carriers) and healthy controls using selective chemiluminigenic substrates that allow differential measurement of the activities of NADPH oxidase and MPO using lucigenin- and luminol-amplified chemiluminescence, respectively [13]. We found a significant increase in MPO-dependent oxidant formation in both CF children and their parents as compared to controls, whereas NADPH activity was similar to that in controls. This increased intracellular MPO activity observed in CF homozygotes and heterozygotes was observed at the basal level as well as following phorbol myristate acetate stimulation but was much greater following opsonized zymosan stimulation. Extracellular release of MPO following stimulation with opsonized zymosan as well as total MPO concentration measured by ELISA in CF neutrophil lysates are not different between CF homozygotes or heterozygotes and controls [14].

This hyperproduction of intracellular MPO-derived oxidants can be normalized by the inhibition of the sodium/proton antiport activity using amiloride or EIPA which are pharmacological blockers of the Na^+/H^+ exchange system as well as in a choline buffer which prevents the exchange system working. The Na^+/H^+ exchangers participate in phagocytosis and subsequent bacterial killing by regulating intracellular pH by compensating for the H^+ load generated by the respiratory burst [15]. Exposure of neutrophils to the phagocytic stimulus opsonized zymosan induces a transient cytosolic acidification followed by a prolonged alkalinization. The latter is inhibited by amiloride analogs and therefore is due to activation of Na^+/H^+ exchange of the NHE-1 type in neutrophils [16]. Thus, our data suggest that CF neutrophils possess altered intracellular pH regulation that may contribute to the observed increased intracellular MPO enzymatic activity.

CF heterozygotes carry the same genetic defect as CF homozygotes but are free from the clinical problems associated with the disorder. The fact that both CF mothers and fathers show the same cellular disturbance as CF children further supports the hypothesis that increased MPO-dependent activity might be linked to the genetic defect. Previous work on physiological abnormalities in CF heterozygotes (reviewed in reference 17) has shown an abnormal sweat test intermediate between those of CF children and controls. It has been proposed that heterozygosity of the CFTR allele has been selected for in certain populations because it protects against typhoid [18]. A different starting point of clinical research in CF treatment has led to the proposal that amiloride could improve the clinical status of CF. The administration of amiloride aerosols to CF patients is based on its Na^+ channel-blocking properties at the level of the airway epithelia [19]. Amiloride treatment enhances water secretion from the cells, increases mucus hydration, improves mucociliary clearance, and delays the clinical decline in lung function. It is tempting to speculate that the beneficial effect of amiloride might be due to its corrective effect on MPO-dependent oxidant hyperproduction.

14.6
MPO in Bronchial Secretions in CF Patients

MPO is found at high concentrations in sputum of CF patients and the degree of lung damage in patients with CF is more closely correlated with sputum MPO activity than with age, gender, CFTR genotype, or bacterial parameters [20]. Using a spectrophotometric assay [21], we have shown that in vitro stimulated neutrophils from CF patients produce more chloramines in the extracellular medium [14] than do normal neutrophils, prompting us to look for the presence of chloramines in CF bronchial secretions. We have shown that sputum from CF patients contains high concentrations of chloramines, along with high levels of taurine and active MPO [22]. Interestingly, respiratory parameters and a nutritional index positively correlate with chloramine levels, whereas a negative correlation with taurine and MPO is observed. Our findings support a dual role for long-lived oxidants at the site of airway inflammation.

14.7
Influence of Infection on MPO Activity and Evaluation of the In Vivo State of Inflammation in CF

Using a chemiluminescence-based assay developed by Allen and Stevens [23, 24], we investigated the possible influence of bacterial infection on the production of MPO-derived oxidants in whole-blood neutrophils from CF patients. We compared luminol luminescence, as a measure of MPO activity, triggered by opsonized zymosan in the absence or in the presence of PAF, a priming agent to further mobilize intracellular reserve pools of opsonin receptors. The increased MPO-dependent oxidant production in isolated neutrophils was greatest following stimulation with complement and immunoglobulin-opsonized zymosan, suggesting a possible linkage to opsonin receptor-mediated activation that could be influenced by infection. MPO-derived oxidant formation triggered by complement-opsonized zymosan, thought to reflect MPO activity in the circulation, is significantly higher in noninfected CF children as compared to healthy children. Moreover, infection with *S. aureus* significantly increases MPO-dependent oxidant formation that is even greater in *P. aeruginosa*-infected CF children. In contrast, parallel measurement of CR1 (CD35) and CR3 (CD11b) expression by flow cytometry did not show any upregulation of these receptors, thus suggesting that this increase luminol-dependent luminescence is due to MPO activity rather than upregulation of opsonin receptors [25].

In vitro PAF priming of whole-blood neutrophils increased intracellular MPO activity as well as surface expression of opsonin receptors as judged by flow cytometry. This allowed us to assess the theoretical reserve of opsonin receptors using the ratio between levels on circulating and PAF-primed cells. The receptor reserve was significantly decreased in all groups of CF patients in comparison to controls and was the lowest in CF children infected with *P. aeruginosa*. Sequential measurement of this receptor reserve during antibiotic treatment in CF children chronically infected with *P. aeruginosa* allowed us to monitor the course of infection with a return to normal levels after eradication of the bacteria. Most interestingly, there was a correlation between this inflammatory index, measured in whole blood CF neutrophils, and the improvement in lung function evaluated by the forced expiratory volume [25].

14.8 Summary

14.8.1 Presence and Role of CFTR in Neutrophils?

Several working hypotheses state that CFTR could be a multifunctional membrane protein involved in controlling channels or exchanges of ions other than Cl^-. But the question of whether a disturbance in Na^+/H^+ antiport in neutrophils from CF homozygotes or heterozygotes is attributable to a defect in CFTR function has not been addressed. Indeed, a defective acidification of intracellular organelles has been shown in nasal polyp cells isolated from CF children [26]. Although the expression of CFTR in cell types of non-epithelial origin and in lymphocytes has been previously described, the presence of CFTR RNA in mature neutrophils has been reported only once [27]. We have been unable to detect CFTR protein in neutrophils using immunoblotting of isolated plasma membranes (unpublished data).

14.8.2 Hypothesis of a Disturbed Control of the Inflammatory Process

Whereas pathogens in the respiratory tract trigger the host inflammatory response, the extent of the response does not necessarily parallel the burden of pathogens. The hypothesis that CF neutrophils have a propensity to produce MPO-dependent oxidants constitutively could change the classical picture of CF inflammation. Rather than representing a single component in host defense against infectious agents, neutrophils would be major contributors to proinflammatory events, being genetically predisposed to release their inflammatory mediators. A recent study has focusing on the analysis of adhesion molecule expression in neutrophils from CF has pointed out a defective shedding of L-selectin after stimulation, thus suggesting a defective control of inflammation in CF [28]. Nonetheless, the molecular basis of neutrophil persistence and predominance in the airways of CF patients has not been explained.

14.8.3 Relationships with MPO Function?

In this model, investigation of MPO-derived oxidant formation appears to be relevant to the pathogenesis of CF as well as important in the monitoring of the inflammatory/infectious state which is crucial to the outcome of CF patients. Our finding showing an increase in MPO-dependent oxidant generation might also be considered in light of recent considerations concerning the function of CFTR as a member of the ATP binding cassette (ABC) protein family [29]. It has been demonstrated that CFTR over-expression decreases intracellular pH [30] and glutathione transport [31]. In this setting, CF neutrophils would show a defect in intracellular glutathione transport and as such increase their intracellular levels of chlorinated oxidants.

Acknowledgements. This work was supported by grants from Association Francaise de Lutte contre la Mucoviscidose (AFLM) and l'Association pour l'Aide a la Recherche contre la Mucoviscidose et l' Assistance aux Malades.

References

1. Boat TF, Welsh MJ, Beaudet AL (1989) Cystic fibrosis. In: Scriver R, Beaudet AL, Ly WS, Valle D (eds) The metabolic basis of inherited disease. McGraw-Hill, New York, pp 2649–2680
2. Stern R (1997) The diagnosis of cystic fibrosis. New Engl J Med 336: 487–491
3. Riordan JR, Rommens JM, Kerem B, Alon N, Rozmahel R, Grzelczak Z, Zielenski J, Lok S, Plavsic N, Chou JL, Drumm ML, Iannuzzi MC, Collins FS, Tsui LC (1989) Identification of the cystic fibrosis gene: cloning and characterization of complementary DNA. Science 245: 1066–1073
4. Welsh MJ, Smith AE (1993) Molecular mechanisms of CFTR chloride channel dysfunction in cystic fibrosis. Cell 73: 1251–1254
5. Konstan MW, Byard PJ, Hoppel CL, Davis PB (1995) Effect of high-dose ibuprofen in patients with cystic fibrosis. N Engl J Med 332: 848–854
6. Khan TZ, Wagner JS, Bost T, Martinez J, Accurso FJ, Riches DWH (1995) Early pulmonary inflammation in infants with cystic fibrosis. Am Respir Crit Care Med 151: 1075–1084
7. Baltimore RS, Mitchell M (1980) Immunologic investigations of mucoid strains of *Pseudomonas aeruginosa*: comparison of susceptibility to opsonic antibody in mucoid and non-mucoid strains. J Infect Dis 141: 238–247
8. Richman-Eisenstat JB, Jorens PG, Hebert CA, Ueki I, Nadel JA (1993) Interleukin-8: an important chemoattractant in sputum of patients with chronic inflammatory airway diseases. Am J Physiol 264: L413–L418
9. Suter S, Schaad UB, Roux L, Nydegger UE, Waldvogel FE (1984) Granulocyte neutral proteases and *Pseudomonas* elastase as possible causes of airway damage in patients with cystic fibrosis. J Infect Dis 149: 523–531
10. Witko-Sarsat V, Halbwachs-Mecarelli L, Schuster A, Nusbaum P, Ueki I, Canteloup S, Lenoir G, Descamps-Latscha B, Nadel JA (1999) Proteinase 3, a potent secretagogue in airways, is present in cystic fibrosis sputum. Am J Respir Cell Mol Biol 20: 729–736
11. Goldman MJ, Anderson GM, Stolzenberg ED, Kari UP, Zasloff M, Wilson JM (1997) Human beta-defensin-1 is a salt sensitive antibiotic in lung that is inactivated in cystic fibrosis. Cell 88: 553–560
12. Brown RK, Kelly FJ (1994) Role of free radicals in the pathogenesis of cystic fibrosis. Thorax 49: 738–742
13. Allen RC (1986) Phagocytic leukocyte oxygenation activities and chemiluminescence: a kinetic approach to analysis. Methods Enzymol 133: 449–493
14. Witko-Sarsat V, Allen RC, Paulais M, Bessou G, Lenoir G, Descamps-Latscha B (1996) Disturbed myeloperoxidase-dependent activity of neutrophils in cystic fibrosis homozygotes and heterozygotes, and its correction by amiloride. J Immunol 157: 2728–2735

15. Grinstein S, Furuya W, Biggar WD (1986) Cytoplasmic pH regulation in normal and abnormal neutrophils. J Biol Chem 261: 512–514
16. Fukushima T, Waddell TK, Grinstein S, Goss GG, Orlowski J, Downey G.(1996) Na^+/H^+ exchange activity during phagocytosis in human neutrophils: role of Fc gamma receptors and tyrosine kinases. J Cell Biol 132: 1037–1052
17. Prince A (1998) The CFTR advantage – capitalizing on a quirk of fate. Nat Med 4: 663–664
18. Gabriel SE, Brigman KN, Koller BH, Boucher RC, Stutts MJ (1994) Cystic fibrosis heterozygote resistance to cholera toxin in the cystic fibrosis mouse model. Science 266: 107–109
19. Knowles MR, Church NL, Waltner WE, Yankaskas JR, Gilligan P, King M, Edwards LJ, Helms RW, Boucher RC (1990) A pilot study of aerosolized amiloride for the treatment of lung disease in cystic fibrosis. N Engl J Med 322: 1189–1194
20. Regelmann WE, Siefferman CM, Herron JM, Elliott GR, Clawson CC, Gray BH (1995) Sputum peroxidase activity correlates with the severity of lung disease in cystic fibrosis. Pediatr Pulmonol 19: 1–9
21. Witko V, Nguyen AT, Descamps-Latscha B (1992) Microtiter plate assay for phagocyte-derived taurine-chloramines. J Clin Lab Anal 6: 47–53
22. Witko-Sarsat V, Delacourt C, Rabier D, Bardet J, Nguyen AT, Descamps-Latscha B (1995) Neutrophil-derived long-lived oxidants in cystic fibrosis sputum. Am J Respir Crit Care Med 152: 1910–1916
23. Allen RC, Stevens DL (1992) The circulating phagocyte reflects the in vivo state of immune defense. Curr Opin Infect Dis 5: 389–398
24. Stevens DL, Bryant AE, Huffman J, Thompson K, Allen RC (1994) Analysis of circulating phagocyte activity measured by whole blood luminescence: correlations with clinical status. J Infect Dis 170: 1463–1472
25. Witko-Sarsat V, Halbwachs-Mecarelli L, Sermet-Gaudelus I, Bessou G, Lenoir G, Allen RC, Descamps-Latscha B (1999) Priming of blood neutrophils in cystic fibrosis children: correlation between functional and phenotypic expression of opsonin receptors pre- and post-PAF priming. J Infect Dis 179: 151–162
26. Barasch J, Kiss B, Prince A, Saiman L, Gruenert D, Al-Awqati Q (1991) Defective acidification of intracellular organelles in cystic fibrosis. Nature 352: 70–73
27. Yoshimura K, Nakamura H, Trapnell BC, Chu CS, Dalemans W, Pavirani A, Lecocq JP, Crystal RG (1991) Expression of the cystic fibrosis transmembrane conductance regulator gene in cells of non-epithelial origin. Nucleic Acids Res 19: 5417–5423
28. Russell KJ, McRedmond J, Mukherji N, Costello C, Keatings V, Linnane S, Henry M, Fitzgerald MX, O'Connor CM (1998) Neutrophil adhesion molecule surface expression and responsiveness in cystic fibrosis. Am J Respir Crit Care Med 157: 756–761
29. Hoof T, Demmer A, Hadam MR, Riordan JR, Tummler BT (1994) Cystic fibrosis-type mutational analysis in the ATP-binding cassette transporter signature of human P-glycoprotein MDR1. J Biol Chem 269: 20575–20583
30 Wei LY, Hoffman MM, Roepe PD (1997) Altered pH regulation in 3T3/CFTR clones and their chemotherapeutic drug-selected derivatives. Am J Physiol 272: C1642-C1653
31. Linsdell P, Hanrahan JW (1998) Glutathione permeability of CFTR. Am J Physiol 275: C323–326

Structure Function Relationships Amongst Members of the Animal Peroxidase Family of Proteins

Rosa Pia Ferrari, Silvio Traversa

15.1 Introduction

Heme peroxidases contain the well-known heme *b* as prosthetic group and use hydrogen peroxide as electron acceptor to catalyze the oxidation of a large number of substrates [1]. On the basis of sequence homology they can be categorized into two superfamilies (I and II) and are present in (I) bacteria, fungi, plants, and (II) animals [2, 3]. The superfamily of animal peroxidases includes myeloperoxidase (MPO), eosinophil peroxidase (EPO), lactoperoxidase (LPO), thyroid peroxidase (TPO), and prostaglandin-endoperoxide synthase (PGHS). These proteins are related members of the same gene superfamily (Table 15.1). The phylogenetic tree of these related members of animal peroxidases, as derived from ProtoMap [4] on the basis of known sequences, predicts some distinguishing divergences among members of the superfamily (Table 15.2).

Evolutionary genetic differentiation of TPO and PGHS might reflect the specialized and distinct physiological functions each has. Cals et al. were the first to provide evidence for expression of the gene encoding LPO in lactating mammary epithelial cells [5]. The phylogenetic tree is based on the organization of structural domains predicted from primary sequences and shared by all members of the family [6]. We restrict our comments in this chapter to LPO, EPO, and MPO, all haloperoxidases by virtue of their in vivo ability to oxidize halides (Br^- and Cl^-, EPO and MPO, respectively) and pseudohalides (SCN^-, LPO). These animal peroxidases oxidize inorganic anions via a bielectronic reaction and, like other peroxidases, can oxidize organic substrates in two monoelectronic steps (Fig. 15.1).

Table 15.1. Animal heme peroxidases

Enzyme	Official name	Common names
1.11.1.7	Peroxidase[a]	Eosinophil peroxidase, lactoperoxidase, myeloperoxidase
1.11.1.8	Iodide peroxidase	Iodinase, iodotyrosine deiodase, thyroid peroxidase
1.14.99.1	Prostaglandin-endoperoxide synthase	Cyclooxygenase, prostaglandin G/H synthase, prostaglandin synthase

[a] Peroxidase (EC 1.11.1.7) includes animal, fungal, plant and bacterial heme peroxidases

Table 15.2. Distinguishing features of members of the animal peroxidase family

Enzyme (EC no.)	Subcellular location	Function
MPO (1.11.1.7)	Azurophilic granules of neutrophils and monocytes	Microbicidal
EPO (1.11.1.7)	Granules of eosinophils	Microbicidal
LPO (1.11.1.7)	Extracellular in mammary and salivary glands	Bacteriostatic and bactericidal
TPO (1.11.1.8)	Membrane of thyroid cells	Thyroid hormone synthesis
PGHS (1.14.99.1)	Microsomal membranes	Synthesis of thromboxanes and prostaglandins

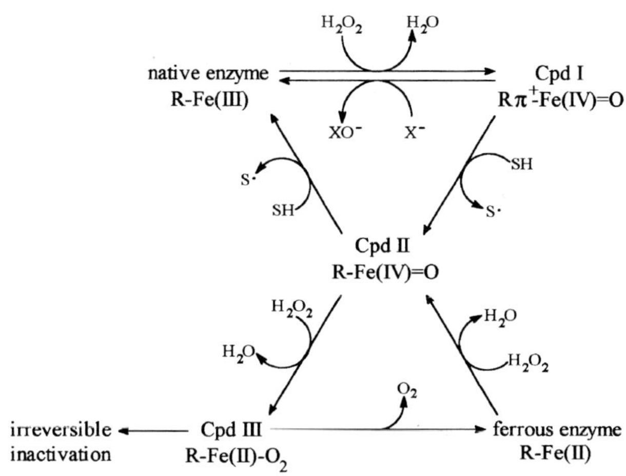

Fig. 15.1. Catalytic mechanism of animal peroxidases

15.2 Sequence Alignment and Homology Degree (EC 1.11.1.7 and EC 1.11.1.8)

We compared the structural organization of the animal peroxidases, aligning the amino acid sequences of proteins derived from the SWISS-PROT databank. Multiple sequence alignment of human MPO, EPO, TPO, and human LPO [7] reveals 23.8% of perfect matches, 13.3% of high similarity and 7.2% of low similarity, respectively. Kooter et al. [8] have reported a 61%, 70%, and 47% identity between MPO and LPO, EPO, and TPO, respectively. We have shown (BLOSUM62 comparison matrix) that bovine LPO shares, respectively, 52.2%, 51.3%, and 47.3% identical residues with human MPO, human EPO, and human TPO, and human LPO shows 51.7%, 51.6% and 47.5% identical residues with human MPO, EPO and TPO, respectively. Further, the sequence of bovine LPO aligned together with the incomplete primary sequence of human LPO yields a 83.3% similarity in the 324 residue overlap.

15.3
Three-Dimensional Structure of MPO, LPO and EPO

The 3-D X-ray structures of canine as well as human MPO [9, 10] and the 3-D theoretical models of bovine LPO and human EPO [11] are currently available. MPO is a homodimer, each subunit being composed of two polypeptides (light chain 13,000 Da, heavy chain 53,500 Da) resulting from post-translational excision of six residues from the common precursor. A single disulfide bridge links each pair of monomers to form the dimeric mature protein. EPO likewise contains two subunits (light chain 15,500 Da and heavy chain 50,000 Da) arranged in a pattern resembling that of hemiMPO but existing as a monomer [12]. Finally, LPO represents the simplest organization and exists as a single-chain polypeptide (78,500 Da) [13]. Because the haloperoxidases exhibit a high degree of structural similarity, the theoretical 3-D models of bovine LPO and human EPO were modeled on the solved X-ray structure of MPO. Four primary potential glycosylation sites were hypothesized for human MPO, bovine LPO and human EPO, and a Ca^{2+} binding site was conserved in all three enzymes.

Very recently the carbohydrate composition and the disulfide bond linkages of bovine LPO have been determined by mass spectrometry [14].

15.4
Heme Active Site Structure of MPO, LPO and EPO

The imidazole ring of proximal histidine (human MPO, His502; bovine LPO, His468; human EPO, His474) lies almost perpendicular to the porphyrin plane and forms a hydrogen bond between the Nδ and an asparagine residue nearby; on the heme distal site, a histidine (human MPO, His261; bovine LPO, His226; human EPO, His233) and an arginine residue (human MPO, Arg405; bovine LPO, Arg372; human EPO, Arg377) are also present and play a crucial role during substrate and cosubstrate binding. These features are shared with the non-homologous plant, fungal and bacterial peroxidases [15]. However, one structural aspect peculiar to animal peroxidases is the covalent linkage of the porphyrin ring to the apoprotein via two ester bonds. The refined X-ray structure of human MPO [10] confirms that the heme prosthetic group is a derivative of protoporphyrin IX in which modified methyl groups on the pyrrole A and C rings form ester linkages with Glu408 and Asp260 respectively. In addition, the vinyl group on ring A is bound covalently with Met409, forming a vinyl sulfonium ion. Recently some authors [8] have hypothesized the formation of a bond between the methionine sulfur atom and the α-carbon of the vinyl group as an alternative to the presence of a sulfonium ion. Kooter et al. have investigated the effect of a Met409Gln mutation on the spectroscopic properties and catalytic activity of MPO [16]. The mutant enzyme exhibits a blue shift of the Soret band, making it similar to that of LPO and EPO, and is unable to peroxidize chloride to hypochlorous acid. This demonstrates that the covalent binding of Met409 with the prosthetic group is essential for the chlorinating activity of MPO.

The structure and mode of binding of the heme group in LPO [11, 17] and EPO [11] have been elucidated recently and are consistent with derivation from protoporphyrin IX, where modified methyl groups on rings A and C form ester bonds with Glu375

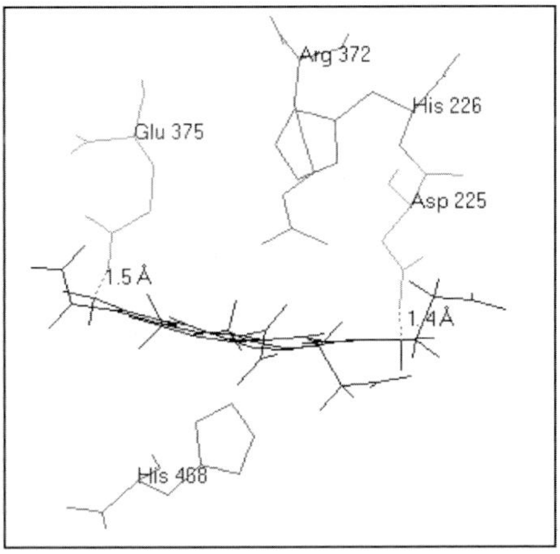

Fig. 15.2. LPO active site (theoretical model [11])

and Asp225. These residues are represented in the active site picture (Fig. 15.2), along with proximal, distal histidines (His468 and His226, respectively) and distal arginine (Arg372), which are involved in the catalytic processing of H_2O_2. Nichol et al. [18] have proposed a disulfide bridge between the protoporphyrin ring and the apoLPO, but our theoretical model suggests that the favorite Cys residue is too distant from heme edges to form the linkage without an unlikely distortion of the overall protein structure. Recently, it has been reported that the reaction of the heme apoLPO complex with hydrogen peroxide results in both autocatalytic modification of the heme and covalent attachment to the protein [19]. This study, carried out on recombinant LPO, has shown that covalent heme attachment may influence the redox potential, substrate binding, and catalytic activity. The heme attachment in MPO, EPO, and TPO is likely to be mediated by a similar mechanism.

15.5
LPO Heterogeneity: Purification and Characterization

Generally, the purification and characterization of LPO leads to heterogeneous samples due to the presence of other proteins, mainly lactoferrin and, sometimes, hemoglobin. The usual tools employed to check this kind of heterogeneity are SDS-PAGE and IEF electrophoresis and these, together with EPR spectroscopy, have been successfully used to test LPO sample purity [13]. The EPR spectrum of LPO samples clearly shows the presence of lactoferrin by presenting a peak at $g = 4.3$ (non-heme iron), and a minor peak at $g = 5.86$ between the main peaks of lactoperoxidase ($g_x = 6.2$; $g_y = 5.4$) probably due to methemoglobin. Figure 15.3 shows the EPR spectrum of highly purified LPO. Even in the absence of other protein impurities, LPO samples can be heterogeneous at the N-terminal sequence. Sequence analysis of cDNA [20] and our own studies (Traversa et al., submitted for publication) have demonstrated that the sequence of mature LPO should begin with Leu129, whereas normally the observed sequence starts with position Asp101 (Traversa et al., submitted for publication; [21]) or sometimes with Ala116, Ser118, or others. These sequence

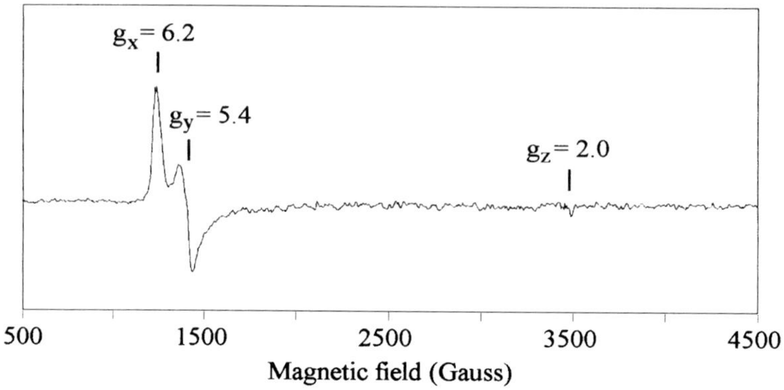

Fig. 15.3. EPR spectrum of LPO

data have been obtained from LPO from bovine milk and colostrum derived from a number of animals. The differences in observed LPO sequences may result in differential structural stability among species. In support of this explanation (Traversa et al., submitted for publication), human LPO has been shown to be more stable in human intestine than is bovine LPO [22]. Recently, polymorphism (two alleles) in the bovine LPO gene (chromosome 19) has been detected and mapped by Karall et al. [23].

15.6 LPO Substrate Binding Studies

LPO plays a key role in the non-immune defense system against bacteria and viruses in milk, saliva, and tears by endogenous two-electron oxidation of thiocyanate to hypothiocyanite. In order to clarify some aspects of the interaction between LPO and its natural substrates (SCN^-, I^-, Br^-), binding studies with several inorganic species (SCN^-, I^-, Br^-, Cl^-, F^-, NO_2^-, N_3^-, CN^-) have been performed using UV-visible spectroscopy [24]. Empirical results were checked by means of computer-assisted simulation of the interaction of such species with the LPO active site structural model [11]. These ligands interact with the protein to form 1:1 complexes and can be assigned to three different groups: (1) SCN^-, I^-, Br^-, and Cl^- (K_D increases along the series); (2) F^- (which shows a singular behavior); and (3) NO_2^-, N_3^-, and CN^- (that bind at the iron site). Computer-assisted docking simulations have shown that all the inorganic substrates examined have the proper steric features to penetrate the active site of LPO. Investigations on the binding of the natural substrates SCN^- and I^- at various pH values and temperatures have revealed that their interaction with LPO involves the protonation of a common site in proximity to the heme iron (possibly distal histidine). Binding studies, EPR spectrum, and simple geometrical considerations concerning the active site structure have allowed us to hypothesize that SCN^- might act as a bridge ligand between His226 and heme iron.

In addition, LPO catalyzes in vitro the oxidation of organic substrates such as phenols, catechol(amine)s, and aromatic amines. In experimental models this enzyme can induce in vivo peroxidation of some carcinogenic substrates to DNA-reactive species. In this context, LPO seems to have a role in inflammatory diseases and in

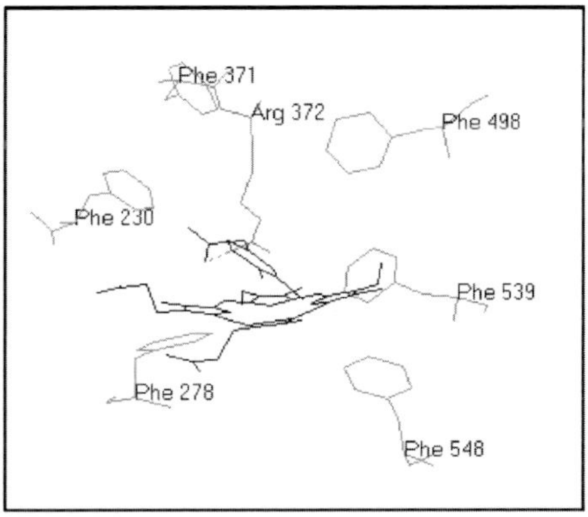

Fig. 15.4. Binding of 3,4-dihydroxyphenylacetic acid to the organic substrate access channel of LPO [25]

mammary carcinogenesis and the interaction between the enzyme and potential organic substrates of biological relevance such as catechol(amine)s have been thoroughly investigated by means of experimental procedures and docking and molecular dynamic simulations [25]. We have demonstrated that catecholic ligands can behave as good structural probes of the heme active site. In particular, the nature of the side terminal chain discriminates the entrance into the protein access channel, whereas side-chain length modulates the interaction with the distal Arg372. In the organic substrate access channel picture shown in Fig. 15.4 a hydrophobic patch formed by six Phe residues (Phe230, Phe278, Phe371, Phe498, Phe539, and Phe548), which are responsible for the hydrophobic interaction with substrate aromatic rings, is clearly discernible. Figure 15.4 also illustrates the interaction between the positively charged Arg372 and the terminal carboxylate of a molecule of 3,4-dihydroxyphenylacetic acid (docking simulation). Further binding and docking studies on a series of biologically relevant organic substrates are in progress in our laboratories in order to design enzyme inhibitor models and to avoid the oxidative activation of dangerous carcinogenic molecules.

15.7 Concluding Remarks

MPO, EPO, LPO (EC 1.11.1.7), and TPO (EC 1.11.1.8) differ physiologically and exhibit distinct catalytic activities, reflecting significant structural variations at the gene and protein levels. In the context of the high homology of primary and overall 3-D structure among multigene family members MPO, EPO, and LPO, the partially different functions seem to be modulated by peculiar structural modifications in the environment of the active site. Since LPO exhibits antibacterial activity and can be easily purified in high quantities, it is arousing increasing interest in the pharmaceutical and medical veterinary fields. Further experiments with these native (or recombinant) and mutated enzymes should better elucidate the structure-function relationships among members of this multigene family of animal peroxidases.

15.8 Summary

Animal peroxidases, namely MPO, EPO, LPO, TPO, and PGHS, are evolutionarily related members of the same gene superfamily. Peroxidases catalyze the peroxidation of halides and pseudohalides via a bielectronic reaction. They play a key role in the antibacterial and innate immune system. Iodide peroxidases induce oxidation of iodide and are involved in the biosynthesis of thyroxine and triiodothyronine. PGHS catalyzes the formation of prostaglandin from arachidonic acid. The similarities in the in vivo functions of MPO, LPO and EPO reflect the striking homology in their primary structures and the presence of common ester linkages between the heme group and the apoprotein. Furthermore, these enzymes seem to induce in vitro catalytic oxidation of organic substrates (two monoelectronic steps) which are regarded as risk factors for inflammatory diseases and possibly breast cancer. MPO, LPO, and EPO are the most studied members of the family. As well as the primary sequences, the X-ray structure of MPO is presently available and a theoretical three-dimensional model has been proposed for both LPO and EPO. In this context peculiar aspects of LPO purification and enzyme-binding interaction have also been analyzed.

References

1. Everse J, Everse KE, Grisham MB (eds) (1991) Peroxidases in chemistry and biology. CRC Press, Boca Raton
2. Welinder KG (1992) Superfamily of plant, fungal and bacterial peroxidases. Curr Opin Struct Biol 2: 388–393
3. Kimura S, Ikeda-Saito M (1988) Human myeloperoxidase and thyroid peroxidase, two enzymes with separate and distinct physiological functions, are evolutionarily related members of the same gene family. Proteins 3: 113–120
4. Yona G, Linial N, Tishby N, Linial M (1998) ProtoMap. http://www.protomap.cs.huji.ac.il
5. Cals MM, Guillomot M, Martin P (1994) The gene encoding lactoperoxidase is expressed in epithelial cells of the goat lactating mammary gland. Cell Mol Biol 40: 1143–1150
6. Corpet F, Gouzy J, Kahn D (1998) The ProDom database of protein domain families. Nucleic Acids Res 26: 323–326
7. Ueda T, Sakamaki K, Kuroki T, Yano I, Nagata S (1997) Molecular cloning and characterization of the chromosomal gene for human lactoperoxidase. Eur J Biochem 243: 32–41
8. Kooter IM, Pierik AJ, Merkx M, Averill BA, Moguilevsky N, Bollen A, Wever R (1997) Difference Fourier transform infrared evidence for ester bonds linking the heme group in myeloperoxidase, lactoperoxidase, and eosinophil peroxidase. J Am Chem Soc 119: 11542–11543
9. Zeng J, Fenna RE (1992) X-ray crystal structure of canine myeloperoxidase at 3 Å resolution. J Mol Biol 226: 185–207
10. Fenna R, Zeng J, Davey C (1995) Structure of the green heme in myeloperoxidase. Arch Biochem Biophys 316: 653–656
11. De Gioia L, Ghibaudi EM, Laurenti E, Salmona M, Ferrari RP (1996) A theoretical three-dimensional model for lactoperoxidase and eosinophil peroxidase, built on the scaffold of the myeloperoxidase X-ray structure. JBIC 1: 476–485
12. Bolscher BGJM, Plat H, Wever R (1984) Some properties of human eosinophil peroxidase, a comparison with other peroxidases. Biochim Biophys Acta 784: 177–186
13. Ferrari RP, Laurenti E, Cecchini PI, Gambino O, Søndergaard I (1995) Spectroscopic investigations on the highly purified lactoperoxidase Fe(III)-heme catalytic site. J Inorg Biochem 58: 109–127
14. Wolf SM, Ferrari RP, Traversa S, Biemann K (2000) Determination of carbohydrate composition and the disulfide bond linkages of bovine lactoperoxidase by mass spectrometry. J Mass Spectrom 35:210–217
15. Dawson JH (1988) Probing structure-function relations in heme-containing oxygenases and peroxidases. Science 240: 433–438

16. Kooter IM, Moguilevsky N, Bollen A, Sijtsema N, Otto C, Wever R (1997) Site-directed mutagenesis of Met243, a residue of myeloperoxidase involved in binding of the prosthetic group. JBIC 2: 191–197
17. Rae TD, Goff HM (1996) Lactoperoxidase heme structure characterized by paramagnetic proton NMR spectroscopy. J Am Chem Soc 118: 2103–2104
18. Nichol AW, Angel LA, Moon T, Clezy PS (1987) Lactoperoxidase haem, an iron-porphyrin thiol. Biochem J 247: 147–150
19. DePillis GD, Ozaki S, Kuo JM, Maltby DA, Ortiz de Montellano PR (1997) Autocatalytic processing of heme by lactoperoxidase produces the native protein-bound prosthetic group. J Biol Chem 272: 8857–8860
20. Dull TJ, Uyeda C, Strosberg AD, Nedwin G, Seilhamer JJ (1990) Molecular cloning of cDNAs encoding bovine and human lactoperoxidase. DNA Cell Biol 9: 499–509
21. Cals MM, Mailliart P, Brignon G, Anglade P, Dumas BR (1991) Primary structure of bovine lactoperoxidase, a fourth member of a mammalian heme peroxidase family. Eur J Biochem 198: 733–739
22. Gothefors L, Marklund S (1975) Lactoperoxidase activity in human milk and in saliva of newborn infants. Infect Immun 11: 1210–1215
23. Karall C, Looft C, Kalm E (1997) Detection and mapping of polymorphisms in the bovine lactoperoxidase (LPO) gene and in the glycosylation-dependent cell adhesion molecule 1 (GlyCAM1) gene using fluorescent single-strand conformation polymorphism analysis. J Anim Sci 75: 1425
24. Ferrari RP, Ghibaudi EM, Traversa S, Laurenti E, De Gioia L, Salmona M (1997) Spectroscopic and binding studies on the interaction of inorganic anions with lactoperoxidase. J Inorg Biochem 68: 17–26
25. Ferrari RP, Traversa S, De Gioia L, Fantucci P, Suriano G, Ghibaudi EM (1998) Catechol(amine)s as probes of lactoperoxidase catalytic site structure: spectroscopic and modelling studies. J Biol Inorg Chem 4: 12–20

Myeloperoxidase (MPO) and Eosinophil Peroxidase (EPO) Deficiency in the Region Friuli-Venezia Giulia of North-Eastern Italy

Pierluigi Patriarca, Maurizio Romano

16.1 Introduction

Our encounter with leukocyte peroxidases occurred during our studies pursuing the nature of the enzyme responsible for the cyanide-insensitive NADPH oxidase activity of stimulated neutrophils. Several candidates, including myeloperoxidase (MPO), were on the scene at that time. In fact, Roberts and Quastel had proposed that the oxidation of NADPH by phagocytosing neutrophils could be accounted for by the oxidase activity of this peroxidase [1]. However, the hypothesis did not gain much support due to insensitivity of the neutrophil NADPH oxidation to cyanide, a known inhibitor of peroxidase activity. But its credit grew when it was reported that the oxidation of NADPH by MPO is not inhibited by cyanide [2]. It was in this context that assessing the NADPH oxidase activity of MPO-deficient neutrophils became almost essential for resolution of the problem.

16.2 MPO and MPO Deficiency

We started searching for MPO-deficient subjects in our region, the Friuli-Venezia Giulia, in collaboration with the clinical pathology laboratories of the hospitals of Pordenone and Udine, the capitals of two of the four provinces of the region. Much to our surprise, a large number [3, 4] of MPO-deficient subjects was identified out of about 100,000 tested, which was an unexpected outcome on the basis of the literature data at that time [5]. It was then possible to undertake several biochemical, morphological, immunological and genetic studies.

Table 16.1 summarizes qualitatively some of the results [3–5]. Incidentally, the NADPH oxidase activity of both resting and phagocytosing MPO-deficient PMN were normal [6]. It soon became clear that the biochemical assay (guaiacol oxidation) usually employed to determine MPO activity was profoundly interfered with by the eosinophil peroxidase (EPO) of even a small proportion (2–3%) of contaminating eosinophils. It was then necessary to find a means to minimize or abolish EPO activity in mixed neutrophil-eosinophil populations. This was achieved in four ways: use of *o*-dianisidine as the substrate [5], selective inhibition of EPO activity with 3-amino-1,2,4-triazole [7], assay of l-alanine decarboxylation in the presence of Triton X-100 and quantitation of MPO peaks in optical absorbance spectra [8]. Figure 16.1 shows how significantly the results of family studies are affected by the biochemical tests used. Several studies, including ours, have indicated an autosomal recessive mode of inheritance of MPO deficiency [9–14]. Very recently Nauseef et al. have sug-

Table 16.1 Some functional properties of MPO-deficient neutrophils

Function	Effect	Reference
Phagocytosis (serum opsonized zymosan)	Same as that of normal cells at early times of contact, then progressively greater	4
Metabolic burst induced by phagocytosis	Same as that of normal cells at early times of contact, then progressively greater	
Metabolic burst induced by phorbol myristate acetate or the chemotactic peptide *N*-formyl-l-norleucyl-l-leucyl-phenylalanine (FNLLP) plus cytochalasin B	Same as that of normal cells	3, 4
Degranulation induced by phagocytosis or FNLLP plus cytochalasin B	Increased for both azurophil and specific granules	3
Bactericidal activity (*E. coli*)	Lower than that of normal cells at early times of contact, normal afterwards	5
Effect of storage	More resistant than normal cells	4

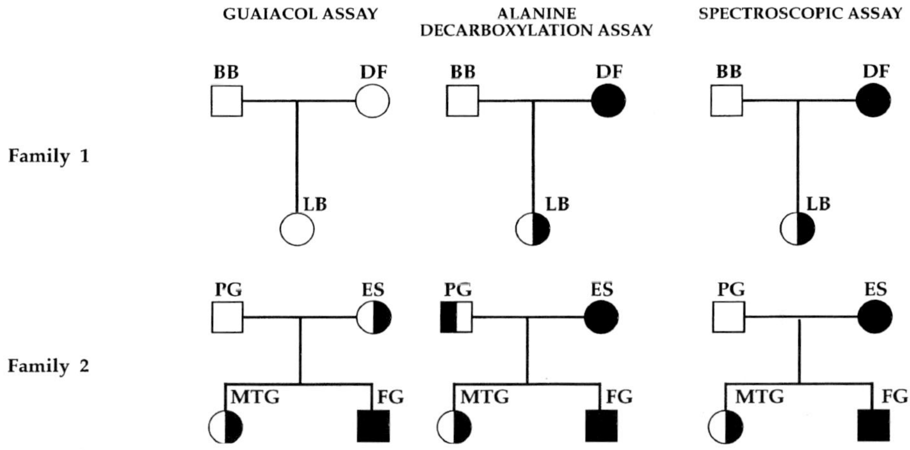

Fig. 16.1. Pedigrees of two MPO-deficient families obtained with three different assays

gested, on the basis of studies on MPO-deficient families in which the propositus bears at least one MPO allele with the R569W missense mutation, that MPO deficiency is frequently the result of compound heterozygosity [15].

Figure 16.2 shows a summary of our molecular studies of an MPO-deficient family [16]. The propositus was a 5-year-old MPO-deficient girl. The father was also MPO-deficient whereas the mother had 24% of normal MPO activity. The father's neutrophils, but not those of the daughter, contained material antigenically related to MPO, while the absorption spectrum of MPO was absent in both subjects. In the MPO gene of the father, two mutations were found, each located on a different allele: the T929C transition within exon 6, causing the nonconservative replacement M251T, and a 14-base deletion within exon 9. The M251T substitution occurred in the carboxyterminal region of the light chain that is included in the heme pocket. The daughter inherited

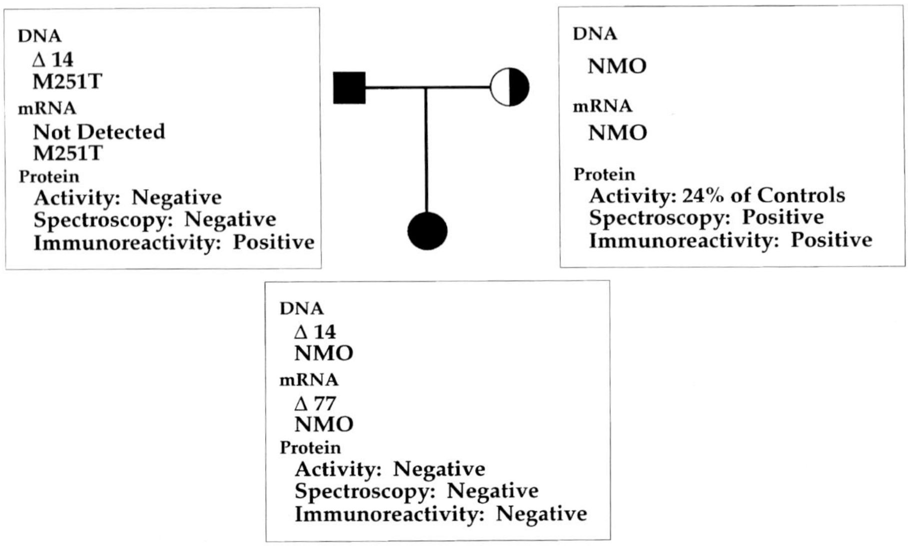

Fig. 16.2. Summary of the results of studies on the molecular characterization of the defect in an MPO-deficient family (*NMO* no mutation observed)

the 14-base deletion from her father. Study of the MPO mRNAs in liquid cultures of granulocyte precursors showed that the 14-base deletion seemed to exhibit different mRNA phenotypes in the father and the daughter. In fact the mRNA derived from the 14-base-deleted allele was not found in the father. Furthermore, an aberrantly spliced MPO mRNA with a 77-base deletion of exon 9, which included the 14-base deletion and led to the generation of a premature stop codon, was found in the daughter. Our data indicate that the alteration of the mRNA context caused by the 14-base deletion provides a basis for the 77-base deletion in the mRNA processing.

Since the granulocyte precursors of the father were more differentiated in the liquid cultures than those of the daughter, the observed different behavior of the 14-base-deleted allele in the father and daughter might have been a result of a differentiation stage-dependent control of altered spliced mRNA. The hypothesis is that the abnormally spliced mRNA may be tolerated during the early stages of differentiation but undergoes degradation during later stages.

In the liquid cultures of the daughter's cells, in addition to the mRNA with the 77-base deletion, a mRNA with the wildtype sequence was also found. This mRNA was inherited from the mother, since no mutations were found in her MPO cDNA or MPO gene. The partial MPO deficiency of the mother might have been caused by as-yet-unidentified regulatory mutation which induces the MPO gene to switch off at an early stage of granulocyte differentiation.

16.3 EPO and EPO Deficiency

The need, mentioned earlier, to distinguish MPO from EPO in mixed neutrophil-eosinophil populations led us to study more in depth the biochemical properties of EPO. The studies then widened out to the biological properties of the molecule. A summary of our results is given in Table 16.2. The effects of EPO on the inflammatory cells (neutrophils and macrophages) are of interest since they prefigure a cytokine-like activity for this molecule. In fact, the activation of neutrophils and macrophages, that is independent of the catalytic activity of EPO, is reminiscent of the effects of two known mediators of inflammation, i.e. TNF-alpha and IFN-alpha. In particular, the macrophage activation in vitro, together with the documented binding of EPO to these cells both in vivo and in vitro, points to a possible role of this molecule in the initiation/maturation of granulomas, particularly the eosinophil-containing granulomas.

Figure 16.3 shows a summary of our studies on the molecular basis of EPO deficiency in one of the five EPO-deficient subjects identified in the clinical pathology laboratory of the Hospital Maggiore in Trieste (out of about 131,000 tested [17]) and his family [18]. The EPO-deficient eosinophils from the propositus contained EPO-related material as determined immunochemically using either monoclonal or polyclonal anti-EPO antibodies, but lacked evidence of spectroscopically detectable EPO. Sequencing of the EPO mRNA extracted from the eosinophil precursors revealed the presence of two mutations. There was a missense mutation, G857A, within exon 7 causing the nonconservative replacement R286H in the heavy chain. In addition there was a G1537 insertion which caused a shift in the reading frame and the appearance of a premature stop codon. The two mutations were located on different alleles, indicating that the resultant defect reflected compound heterozygosity. Both the son and the daughter of the proband had inherited the G857A transition and their eosinophils contained peroxidase activity intermediate between that of control subjects and the proband, suggesting that the R286H causes the functional deficiency. The eosinophil precursors from the EPO-deficient subject were found to actively synthesize an EPO that was apparently normal in terms of cytochemical reaction for peroxidase and immunoreactivity with monoclonal and polyclonal anti-EPO antibodies, but abnor-

Table 16.2. EPO properties

Property	Reference
Catalyzes the bromide-dependent decarboxylation of the apolar amino acid L-alanine	20
3-amino-1,2,4-triazole inhibits EPO, but not MPO activity	7
Catalyzes oxidation of homovanillic acid at highly alkaline pH	19
Binds in vitro to several cell types including neutrophils and macrophages	21, 22
Stimulates neutrophil aggregation and neutrophil adhesion to the unstimulated endothelium	23
Macrophages in the eosinophil granuloma of the Hand-Schüller-Christian disease are decorated with EPO originating from eosinophil degranulation	24
Shows cytokine-like activities, since it potentiates the PMA-triggered metabolic burst of human monocyte-derived macrophages, and stimulates TNF secretion by human monocyte-derived macrophages	25

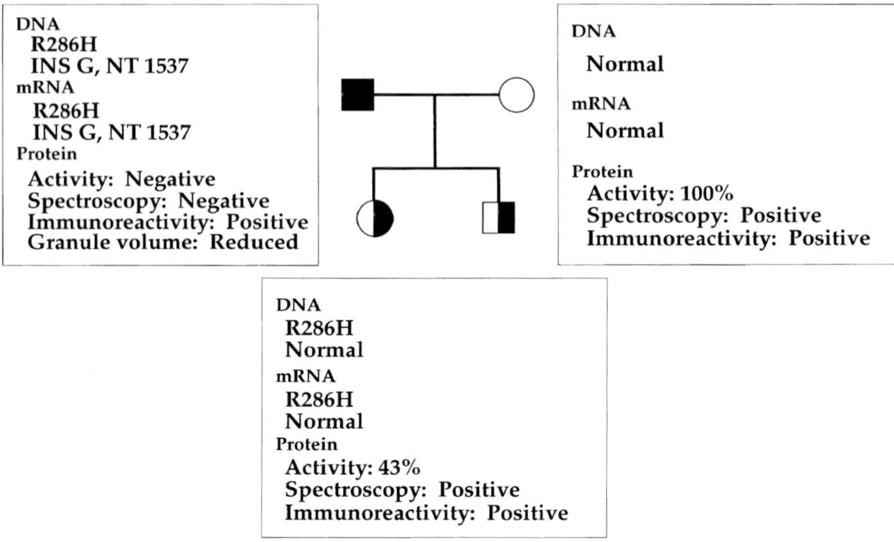

Fig. 16.3. Summary of the results of studies on the molecular characterization of the defect in an EPO-deficient family

mal in terms of spectroscopic properties. The cytochemical reaction for peroxidase progressively disappeared as the eosinophil precursors of the EPO-deficient subject completed the maturation, suggesting that the R286H substitution affects the stability of the enzyme [18].

Even the studies with the EPO-deficient subjects required refinements of the usual technical approaches in order to avoid interference by MPO with the assay of EPO activity in eosinophil populations possibly contaminated with neutrophils. This was achieved by devising selective biochemical assays for EPO, i.e. the bromide-dependent decarboxylation of L-alanine [8] or the oxidation of homovanillic acid at highly alkaline pH [19].

Recently, we have succeeded in producing normal and mutated (R286H) recombinant proEPO as the first step in a series of studies on structure-function relationships. The normal and R286H recombinant EPOs were expressed in the K-562 cell line and the impact of the R286H substitution on the enzymatic activity and the biosynthesis of EPO was analyzed. These studies show that the acquisition of the peroxidase activity occurs at the level of the EPO precursor and the biochemical properties of the recombinant wildtype EPO precursor are comparable to those of the natural enzyme. Although the R286H substitution does not affect the heme insertion, the R286H substitution probably causes an alteration of EPO folding that affects the catalytic properties of the precursor but not its stability.

16.4 Summary

Our interest in MPO originated from our studies on the nature of the enzyme responsible for the neutrophil (PMN) respiratory burst, i.e. NADPH oxidase. It has been proposed that the oxidase activity of MPO may account for PMN NADPH oxidase activity. Testing the NADPH oxidase activity of MPO-deficient PMN would have been the obvious approach to the problem. The search for MPO-deficient subjects we undertook to test that hypothesis revealed a surprisingly high prevalence of the defect in the Friuli-Venezia Giulia region. It was, therefore, possible not only to assay the NADPH oxidase activity of MPO-deficient PMN, that turned out to be normal, but also to extend the studies with biochemical, immunological and genetic approaches. EPO was a disturbing character in our studies on MPO deficiency, since it interfered profoundly with MPO biochemical assays. We therefore sought to elucidate its properties in order to find a means to distinguish EPO and MPO in mixed neutrophil-eosinophil populations. Moreover, biochemical, immunological and genetic studies have been carried out on an EPO-deficient family.

References

1. Roberts J, Quastel J (1964) Role of peroxidase in the oxidation of NADPH by phagocytosing neutrophils. Nature 202: 85–89
2. Takanaka K, O'Brien PJ (1975) Mechanisms of hydrogen peroxide formation in leukocytes: the NAD(P)H oxidase activity of myeloperoxidase. Biochem Biophys Res Commun 62: 971–977
3. Dri P, Cramer R, Menegazzi R, Patriarca P (1985) Increased degranulation of human myeloperoxidase-deficient polymorphonuclear leucocytes. Br J Haematol 59: 115–125
4. Dri P, Soranzo MR, Cramer R, Menegazzi R, Miotti V, Patriarca P (1985) Role of myeloperoxidase in respiratory burst of human polymorphonuclear leukocytes. Studies with myeloperoxidase-deficient subjects. Inflammation 9: 21–31
5. Cramer R, Soranzo MR, Dri P, Rottini GD, Bramezza M, Cirielli S, Patriarca P (1982) Incidence of myeloperoxidase deficiency in an area of northern Italy: histochemical, biochemical and functional studies. Br J Haematol 51: 81–87
6. Patriarca P, Dri P, Kakinuma K, Tedesco F, Rossi F (1975) Studies on the mechanism of metabolic stimulation in polymorphonuclear leucocytes during phagocytosis. I. Evidence for superoxide anion involvement in the oxidation of NADPH2. Biochim Biophys Acta 385: 80–86
7. Cramer R, Soranzo MR, Dri P, Menegazzi R, Pitotti A, Zabucchi G, Patriarca P (1984) A simple reliable assay for myeloperoxidase activity in mixed neutrophil-eosinophil cell suspensions: application to detection of myeloperoxidase deficiency. J Immunol Methods 70: 119–125
8. Dri P, Cramer R, Soranzo MR, Comin A, Miotti V, Patriarca P (1982) New approaches to the detection of myeloperoxidase deficiency. Blood 60: 323–327
9. Larrocha C, Fernandez de Castro M, Fontan G, Viloria A, Fernandez-Chacon JL, Jimenez C (1982) Hereditary myeloperoxidase deficiency: study of 12 cases. Scand J Haematol 29: 389–397
10. Lehrer RI, Cline MJ (1969) Leukocyte myeloperoxidase deficiency and disseminated candidiasis: the role of myeloperoxidase in resistance to *Candida* infection. J Clin Invest 48: 1478–1488
11. Salmon SE, Cline MJ, Schultz J, Lehrer RI (1970) Myeloperoxidase deficiency. Immunologic study of a genetic leukocyte defect. N Engl J Med 282: 250–253
12. Cech P, Papathanassiou A, Boreux G, Roth P, Miescher PA (1979) Hereditary myeloperoxidase deficiency. Blood 53: 403–411
13. Kutter D, Thoma J, al-Haidari K, Trierweiler P (1993) Coexistence of two distinct populations of neutrophilic granulocytes, one normal and one partial MPO-deficient. Acta Clin Belg 48: 401–404
14. Bos AJ, Wever R, Hamers MN, Roos D (1981) Some enzymatic characteristics of eosinophil peroxidase from patients with eosinophilia and from healthy donors. Infect Immun 32: 427–431
15. Nauseef WM, Cogley M, Bock S, Petrides PE (1998) Pattern of inheritance in hereditary myeloperoxidase deficiency associated with the R569W missense mutation. J Leukoc Biol 63: 264–269

16. Romano M, Dri P, Dadalt L, Patriarca P, Baralle FE (1997) Biochemical and molecular characterization of hereditary myeloperoxidase deficiency. Blood 90: 4126–4134
17. Zabucchi G, Soranzo MR, Menegazzi R, Vecchio M, Knowles A, Piccinini C, Spessotto P, Patriarca P (1992) Eosinophil peroxidase deficiency: morphological and immunocytochemical studies of the eosinophil-specific granules. Blood 80: 2903–2910
18. Romano M, Patriarca P, Melo C, Baralle FE, Dri P (1994) Hereditary eosinophil peroxidase deficiency: immunochemical and spectroscopic studies and evidence for a compound heterozygosity of the defect. Proc Natl Acad Sci U S A 91: 12496–12500
19. Menegazzi R, Zabucchi G, Zuccato P, Cramer R, Piccinini C, Patriarca P (1991) Oxidation of homovanillic acid as a selective assay for eosinophil peroxidase in eosinophil peroxidase-myeloperoxidase mixtures and its use in the detection of human eosinophil peroxidase deficiency. J Immunol Methods 137: 55–63
20. Cramer R, Soranzo MR, Patriarca P (1981) Evidence that eosinophils catalyze the bromide-dependent decarboxylation of amino acids. Blood 58: 1112–1118
21. Zabucchi G, Menegazzi R, Soranzo MR, Patriarca P (1986) Uptake of human eosinophil peroxidase by human neutrophils. Am J Pathol 124: 510–518
22. Zabucchi G, Soranzo MR, Menegazzi R, Bertoncin P, Nardon E, Patriarca P (1989) Uptake of human eosinophil peroxidase and myeloperoxidase by cells involved in the inflammatory process. J Histochem Cytochem 37: 499–508
23. Zabucchi G, Menegazzi R, Cramer R, Nardon E, Patriarca P (1990) Mutual influence between eosinophil peroxidase (EPO) and neutrophils: neutrophils reversibly inhibit EPO enzymatic activity and EPO increases neutrophil adhesiveness. Immunology 69: 580–587
24. Zabucchi G, Soranzo MR, Menegazzi R, Cattin L, Vecchio M, Lanza F, Patriarca P (1991) Eosinophilic granuloma of the bone in Hand-Schuller-Christian disease: extensive in vivo eosinophil degranulation and subsequent binding of released eosinophil peroxidase (EPO) to other inflammatory cells. J Pathol 163: 225–231
25. Spessotto P, Dri P, Bulla R, Zabucchi G, Patriarca P (1995) Human eosinophil peroxidase enhances tumor necrosis factor and hydrogen peroxide release by human monocyte-derived macrophages. Eur J Immunol 25: 1366–1373

Chapter 17

Biochemistry and Physiology of Thyroid Peroxidase

Jan J.M. de Vijlder, Hennie Bikker

17.1 Introduction

The thyroid gland is composed of follicles, the smallest entity needed for the production of thyroid hormone, the main function of the thyroid. The follicular lumen, filled up with thyroglobulin, is surrounded by thyrocytes [1]. These thyroid cells are able to produce the proteins and factors needed for thyroid hormone synthesis. Thyroglobulin is the protein in which iodine, bound to tyrosine residues, and thyroid hormone, formed from these iodinated tyrosine residues, is stored until needed. Iodination and coupling of iodo-tyrosine residues is catalyzed by the enzyme thyroid peroxidase (TPO) [2].

17.2 Structure and Function of Thyroid Peroxidase

TPO is a transmembrane, monomeric enzyme, located in the apical cell membrane where the active, heme-containing site protrudes from the cell into the follicular lumen. The human TPO gene spans 150 kb and is located on chromosome 2p25-24. The gene is comprised of 17 exons containing 3-kb coding sequences. TPO is a glycosylated, heme-containing enzyme consisting of 933 amino acids and has a molecular mass of about 100,000 Da. TPO-2, a molecule of a somewhat smaller size, is the translation product of an alternatively spliced TPO mRNA, where exon 10 coding for part of the active site is spliced out [3]. It has been shown that this protein, normally present in thyroid tissue, is enzymatically inactive [4, 5].

TPO shares a high degree of homology with other mammalian peroxidases such as myeloperoxidase (MPO), eosinophil peroxidase (EPO) and lactoperoxidase (LPO). All these peroxidases have a conserved heme environment and a calcium-binding loop in common and likely arose from a common ancestor [6]. TPO shows homology particularly with LPO. Both are monomeric enzymes, but TPO is a membrane bound enzyme, and LPO is not.

The function of TPO is a very specific one, namely catalyzing the iodination of tyrosine residues and the synthesis of thyroid hormone in thyroglobulin. Accordingly TPO, in contrast to the other peroxidases, shows a redox potential by which only iodide and thiocyanate, and not bromide or chloride, can be oxidized. A higher potential would distort its function. This basis for the observed difference in redox potential is not well understood, as the active center in all these peroxidases is well conserved containing the same porphyrin group. Conceivably, the difference in an amino acid at position 243, in the neighborhood of the heme group (Table 17.1), may play a role in this difference.

Table 17.1. Redox potentials of mammalian peroxidases in relation to the amino acid at position 243 (according to Bikker [6]). The redox potentials are based on the redox potentials (in parentheses) of the following reactions: $2I^- \leftrightarrow I_2+2e$ (0.54), $2CNS^- \leftrightarrow (CNS)_2 +2e$ (0.77), $2Br^- \leftrightarrow Br_2+2e$ (1.07), $2Cl^- \leftrightarrow Cl_2+2e$ (1.36), $2H_2O \leftrightarrow H_2O_2+2H^++2e$ (1.77) (data collected from references 11–14)

Enzyme	Able to oxidize	Amino acid at 243	Redox potential (V)[a]
MPO	Cl^-, Br^-, CNS^-, I^-	Methionine	1.36–1.77
EPO	Cl^-, Br^-, CNS^-, I^-	Threonine	1.36–1.77
LPO	Br^-, CNS^-, I^-	Glutamine	1.07–1.36
TPO	CNS^-, I^-	Valine	0.54–1.07

[a] Estimated redox potential = standard oxidation-reduction potential of the couple (XPO-native enzyme ↔ XPO-compound I)+2e [11]

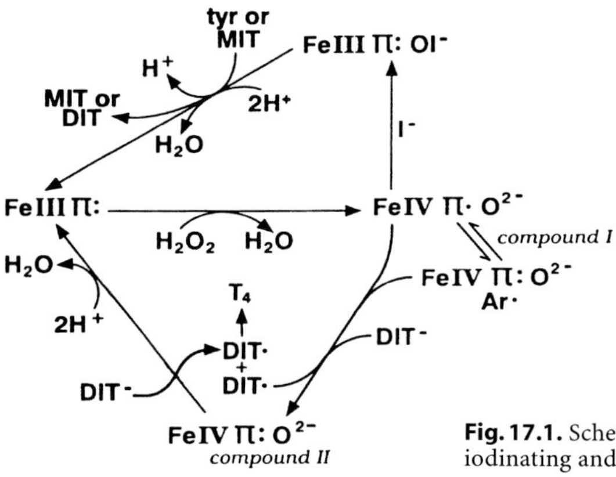

Fig. 17.1. Scheme of the reaction mechanism of the iodinating and coupling activities of thyroid peroxidase

Although it has been known for 50 years that thyroxine, T4, is formed in the thyroid gland from its precursor 3,5-diiodotyrosine, the mechanism of this conversion is still not completely understood. Two possible mechanisms have been proposed and are generally referred to as intramolecular and intermolecular coupling processes. Intramolecular coupling has been proposed by several authors, e.g. Taurog et al. [7]. In this case native TPO is oxidized to compound I, from which two pathways can be distinguished. The iodination that proceeds according to an electrophilic substitution reaction and the coupling, that may be regarded as a two-step reaction in which diiodotyrosine (DIT) radicals are formed which in turn react with each other to form T4 (Fig. 17.1). It has been found that even traces of added DIT stimulate the reaction but do not directly participate.

The intermolecular coupling reaction mechanism was originally proposed by Ogawara and Cahnmann [8]. According to that mechanism, DIT couples readily with its keto analog 3,5-diiodo-4-hydroxyphenyl pyruvic acid (DIHPPA). However, this reaction involves the conversion of a DIT residue in thyroglobulin, for instance, with a free DIHPPA molecule. Recently Oza et al. [9] have proposed a mechanism for the in vitro conversion between the oxidized DIHPPA and free DIT. If this mechanism holds true in vivo, then free DIHPPA has to be incorporated into the thyroxine molecule, a reaction which has not yet been demonstrated.

It is clear that thyroxine synthesis must be facilitated by ionization of the phenoxy groups involved. For that reason investigation of the pH dependence of the coupling reaction would be informative. In comparing the formation of T4 and 3,3'-diiodothyronine (T2) in thyroglobulin, catalyzed by LPO or TPO a strong indication has been obtained that phenoxy ions are indeed required. The half-value for the coupling of two DIT residues or of two MIT residues in thyroglobulin has been found at pH 6.5 and pH >8, respectively. Both observations are in accordance with the pKa values of the hydroxyl group of DIT (6.5) and of MIT (8.7). This observation is strengthened by the finding that the coupling of methoxy phenol (guaiacol) with a pKa value of 7, also shows a half-maximal value at pH 7 [10]. In this in vitro iodothyronine formation system, we have used nonenzymatically iodinated thyroglobulin freed from iodide. The DIT residues present in the peptide chain were alone sufficient for iodothyronine formation. The formation of triiodothyronine (T3) follows a pattern indicating that ionization of the phenoxy group of the DIT residue, becoming the inner ring of the iodothyronine molecule, is more important than ionization of the outer ring. For reversed triiodothyronine (rT3) it is the MIT residue that appears to have the largest influence [10].

17.3 Summary

From these observations we conclude that:

- Ionization of the phenoxy group is required for iodothyronine formation.
- DIT radicals are thought to be involved in intramolecular T4 formation.
- Traces of DIT stimulate the formation of iodothyronine but do not participate in the reaction.

References

1. DeGroot LJ, Stanbury JB (1975) Structure and function of the thyroid. In: The thyroid and its diseases, 4th edn. John Wiley & Sons, New York London Sydney Toronto, pp 1–36
2. Taurog A (1996) Hormone synthesis. In: Braverman LE, Utiger RD (eds) The thyroid. Lippincott-Raven, Philadelphia, pp 47–84
3. De Vijlder JJM, Bikker H, Ris Stalpers C, Vulsma T (1997) Structure, function and relevance of thyroid peroxidase in inherited diseases of the thyroid. Curr Opin Endocrinol Diabetes 4: 328–332
4. Bikker H, Waelkens JJJ, Bravenboer B, De Vijlder JJM (1996) Congenital hypothyroidism caused by a premature termination signal in exon 10 of the human thyroid peroxidase gene. J Clin Endocrinol Metab 81: 2076–2079
5. Niccoli P, Fayadat L, Panneels V, Lanet J, Franc J-L (1997) Human thyroperoxidase in its alternatively spliced form (TPO2) is enzymatically inactive and exhibits changes in intracellular processing and trafficking. J Biol Chem 272: 29487–29492
6. Bikker H (1996) Related peroxidases. In: Thyroid peroxidase. Dysfunction in total iodide organification defects. PhD Thesis, University of Amsterdam
7. Taurog A, Dorris ML, Doerge DR (1996) Mechanism of simultaneous iodination and coupling catalyzed by thyroid peroxidase. Arch Biochem Biophys 334: 284–294
8. Ogawara H, Cahnmann HJ (1972) Nonenzymic synthesis of iodothyronine residues in thyroglobulin. Biochem Biophys Acta 257: 328–338
9. Oza VB, Salamonczyk GM, Guo Z, Sih CJ. (1997) Model reactions of thyroxine biosynthesis. Identification of key intermediates in thyroxine formation from 3,5-diiodo-l-tyrosine and 4-hydroxy-3,5-diiodophenylpyruvic acid. J Am Chem Soc 119: 11315–11316
10. De Vijlder JJM, Den Hartog MT (1998) Anionic iodotyrosine residues are required for iodothyronine synthesis. Eur J Endocrinol 138: 227–231

11. Taurog A, Dorris ML (1991) Peroxidase-catalyzed bromination of tyrosine, thyroglobulin, and bovine serum albumin: comparison of thyroid peroxidase and lactoperoxidase. Arch Biochem Biophys 287: 288–296
12. Aune TM, Thomas EL (1977) Accumulation of hypothiocyanite ion during peroxidase catalyzed oxidation of thiocyanate ion. Eur J Biochem 7: 1297–1306
13. Mayeno AN, Curran J, Roberts RL, Foote CS (1989) Eosinophils preferentially use bromide to generate halogenating agents. J Biol Chem 264: 5660–5668
14. Slungaard A, Mahoney JR (1991) Thiocyanate is the major substrate for eosinophil peroxidase in physiologic fluids. J Biol Chem 266: 4903–4910

Severe Congenital Hypothyroidism Caused by Mutations in the Thyroid Peroxidase Gene

Hennie Bikker, Jan J.M. de Vijlder

18.1 Introduction

Congenital hypothyroidism (CH) occurs relatively frequent in the Netherlands with a prevalence of 1:3000 [1]. Because early detection and prompt therapy are needed to prevent irreversible damage to the central nervous system, a neonatal screening program was developed.

In about 70% of these patients, agenesis or dysgenesis is the cause of CH. Hereditary inborn errors are found in about 15% of the CH patients. In the Netherlands, total iodide organification defects (TIODs), are the most common hereditary inborn error causing CH. The prevalence of TIOD in the Netherlands is 1:60,000. The patients show severe hypothyroidism having low T4 and T3 plasma concentrations at birth. These low plasma concentrations are of maternal origin. TSH and thyroglobulin levels are high, indicating a stimulated, active thyroid gland. One of the most important characteristics is the positive perchlorate discharge test. The administration of sodium perchlorate washes out more than 90% of the trapped radioiodide, demonstrating that the iodide present in the gland is not bound to protein. Under normal conditions iodide taken up by the gland is incorporated in thyroglobulin at the apical membrane of the thyroid cell.

Since organification requires both oxidation and peroxidase activity, abnormalities can be caused by defects in H_2O_2 generation or in thyroid peroxidase (TPO). In thyroid tissues of three patients with TIOD, available for investigation, no peroxidase activity was detectable. Thus, it became clear that the iodide taken up by the thyroid gland is not oxidized and bound to thyroglobulin and that production of thyroid hormones is not possible [2].

18.2 Mutation Analysis of TPO Deficiency

Early detection and treatment of CH prevents the occurrence of goiters in affected individuals. As a consequence, thyroid tissue from these patients is seldom obtained. For this reason we performed diagnosis at the molecular level in genomic DNA obtained from white blood cells [3]. Presently in the Netherlands there are about 45 families known in which TIOD has occurred. Patients from 31 of these families were investigated for mutations in the TPO genes.

In 23 families we detected inactivating mutations in both TPO alleles. Some patients were homozygous for one mutation, and others were compound heterozygous for two different mutations. In seven families, we found an inactivating muta-

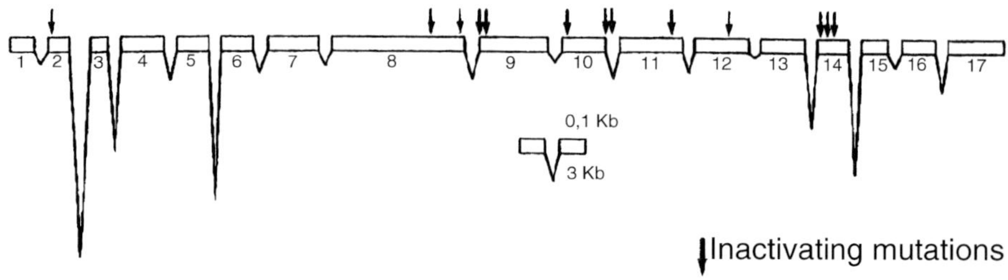

Fig. 18.1. Mutations in the TPO gene cosegregating with TIOD

tion in only one allele. In one family no inactivating mutations have been detected as yet. In this way 13 different mutations, cosegregating with TIOD, were found. In Fig. 18.1 the TPO gene is shown in which these mutations are indicated.

The duplication in exon 8, the T to G single base mutation in exon 9 and the duplication in exon 2 are the mutations that occur most often. Together they account for more than 50% of the mutations found in our patient population. Of the 13 mutations, 6 give a frame shift, followed by an early termination signal. One mutation in exon 10 results in a direct termination signal [4]. The TPO mRNA level in the thyroid gland of this patient was reduced and consisted mainly of the alternatively spliced form of TPO missing exon 10 (TPO-2). Only the enzymatically inactive TPO-2 translation product was detected in thyroid tissue of this patient.

Other mutations altered amino acids. In order to exclude the possibility that these changes were cosegregating polymorphisms, we investigated the impact of these mutations on TPO expression and activity. For this purpose we used the Semliki Forest Virus replication system [5]. We determined the TPO iodination activity and the coupling activity using the iodide assay and guaiacol assay, respectively. Although all mutants were expressed, only recombinant TPO with mutation at position 1858 resulted in enzymatically active TPO comparable to wildtype activity. The other mutations in exon 9 and 14 yielded TPO that was below the detection level of both assays.

The two mutations in exon 9 resulted in the introduction of a phenylalanine (Ile447Phe) or in the loss of a tyrosine (Tyr453Asp). The presence of amino acids containing aryl groups appeared in the case of myeloperoxidase to be very important for the structure. The mutation at the exon 10/intron 10 border (G1858 A) yielded recombinant TPO that was enzymatically active. This mutation was found in three unrelated families with patients with TIOD and could in theory also have affected splicing. In that case exon 10 would be skipped, resulting in inactive TPO-2. No thyroid tissue was available for investigation and the possibility of alternative splicing was not checked.

Since the related peroxidases such as myeloperoxidase contain no membrane-spanning part, the sequence around exon 14 could not be compared to the other peroxidases. But the fact that both tested mutations resulted in completely inactive TPO indicates that this region of the protein is also very important for proper folding of the protein and for enzyme activity.

18.3 Summary

Practically all defects causing TIOD are the result of inactivating mutations in TPO. One-third of the mutations underlying TIOD are caused by a GGCC duplication in exon 8. More than 50% of the TIODs are caused by only three inactivating mutations in TPO. Analysis of mutations inactivating TPO in this expression system demonstrated the importance of specific amino acids for the structure and function of TPO.

The introduction of other mutations – recently discovered mutations in patients or deliberate mutations altering amino acids in the neighborhood of the heme group or the membrane spanning site – will give us more insight into the structure and function of TPO.

References

1. Vulsma T (1991) Etiology and pathogenesis of congenital hypothyroidism. PhD Thesis. University of Amsterdam, The Netherlands
2. Bikker H (1996) Thyroid peroxidase. Dysfunction in total iodide organification defects. PhD Thesis. University of Amsterdam, The Netherlands
3. Bikker H, Vulsma T, Baas F, Vijlder JJM de (1995) Identification of five novel inactivating mutations in the human thyroid peroxidase gene by denaturing gradient gel electrophoresis. Hum Mutat 6: 9–16
4. Bikker H, Waelkens JJJ, Bravenboer B, Vijlder JJM de (1996) Congenital hypothyroidism caused by a premature termination signal in exon 10 of the human thyroid peroxidase gene. J Clin Endocrinol Metab 81: 2076–2079
5. Bikker H, Baas F, Vijlder JJM de (1997) Molecular analysis of mutated thyroid peroxidase detected in patients with total iodide organification defects. J Clin Endocrinol Metab 82: 649–653

Screening for Leukocyte Peroxidase Deficiencies by Means of Flow Cytometry: Application to the Study of Prevalence, Pathology and Genetics

Dolphe Kutter, Luc Verstraeten

19.1 Introduction

To gain an insight into the prevalence, pathology and genetics of peroxidase deficiencies it is necessary to study a large number of cases. This means screening large population groups. In classical haematology the detection of peroxidase deficiencies is only possible by special peroxidase staining techniques that are not commonly used in daily routines. Thus, the first case of myeloperoxidase (MPO) deficiency was discovered by chance by means of peroxidase staining. Descriptions of similar cases remained rare until the introduction of the first Technicon Hemalog Analyser and the more modern versions now produced by Bayer (H-1, H-3, ADVIA). The unique detector system of these instruments automatically combines two-angle laser diffraction of red cells and leukocyte differentiation based on their peroxidase activity. Figure 19.1 shows the work flow of a complete blood count. The "erythrogram" obtained by measuring both the volume and haemoglobin concentration of approximately 50,000 red blood cells allows an insight into red cell morphology. Differentiation of white blood cells is displayed by plotting their size against their peroxidase activity, as measured after exposure to a peroxide/chromogen reagent. In the normal PEROX diagram of these instruments, the white blood cells are clearly separated into five clusters corresponding to neutrophils (N), eosinophils (E), monocytes (M), lymphocytes (L) and LUC (large unstained cells) (Fig. 19.2A). Basophils are differentiated in a separate operation: a sample of blood is treated with an acid buffer, which deprives all the white cells of their protoplasm ("protoplasmic stripping"), leaving the bare nuclei and differentiating round nuclei from more complex ones. Only basophils withstand this treatment and are counted in a separate BASO channel. This allows routine detection not only of complete peroxidase deficiency of certain types of white blood cells but also of all degrees of partial deficiency, with the possibility of quantifying the defect.

19.2 MPO Deficiency

19.2.1 Detection and Quantitation of Neutrophil MPO Deficiency

Normal neutrophils occupy the upper right quadrant of the PEROX diagram (Fig. 19.2A). With decreasing MPO activity the cluster is displaced towards the left. Intermediate activity – partial deficiency – is represented by clusters in the upper middle region (Fig. 19.2B), and clusters very close to or in the LUC correspond to sub-

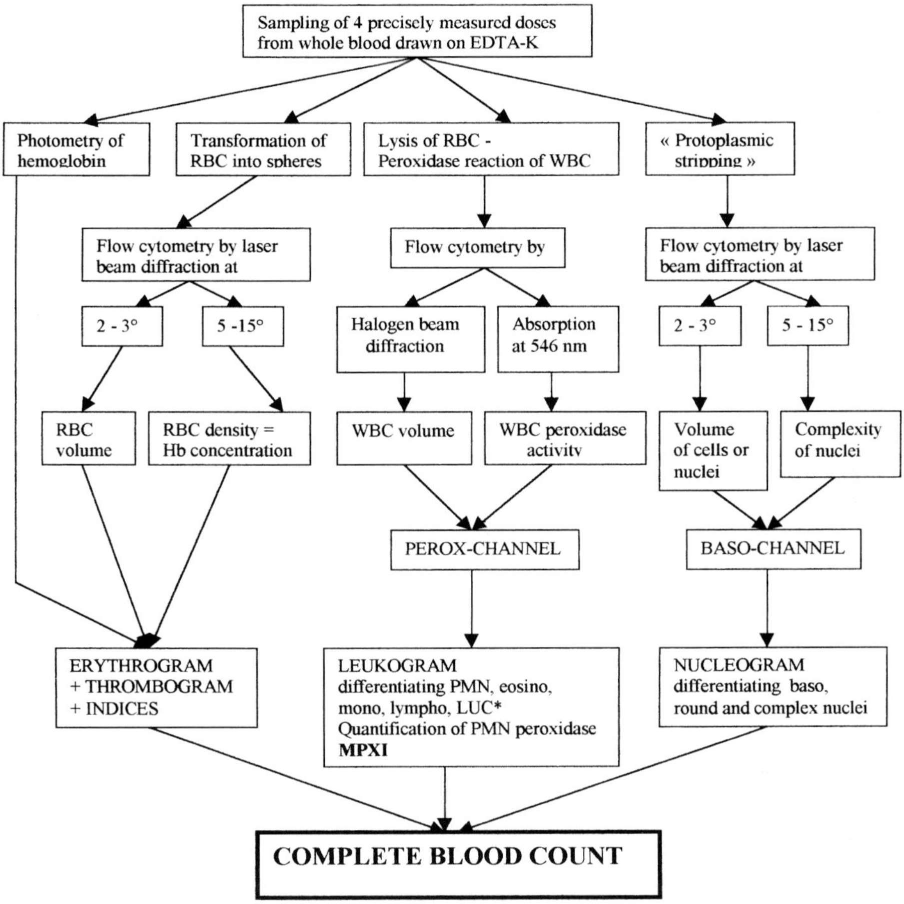

Fig. 19.1. Details of the determination of complete blood count using the Bayer hematological analysers H-1, H-2, H-3 and ADVIA (**LUC* large unstained cells)

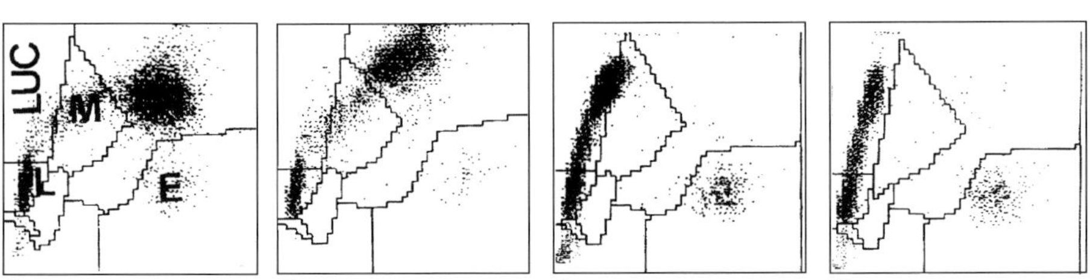

Fig. 19.2A-D. ADVIA displays. **A** normal: *N* neutrophils, *E* eosinophils, *M* monocytes, *L* lymphocytes, *LUC* large unstained cells. **B** Partial MPO deficiency (MPXI –25). **C** Subtotal MPO deficiency (MPXI –56). **D** Total MPO deficiency (MPXI –64)

total or total deficiency (Fig. 19.2C,D). The degree of deficiency is expressed by the MPXI, the MPO index, which the instrument calculates from the coordinates of the neutrophilic cluster, comparing them to ideal coordinates. Values range from +10 to −65. In a large population study on over 41,000 routine patients the MPXI enabled delimitation of several slightly overlapping subpopulations [1]. By far the most important group of approximately 90% of the probands was represented by a Gaussian curve limited by the −10 to +10 values. This confirms the normal values claimed by the producers of the instrument. Another overlapping Gaussian curve approximately limited by 0 and −20 represents mild deficiencies in 9%. Lower values are rare. We consider values between −21 and −50 as partial deficients. With a frequency of about 1% their low number does not permit clear delimitation of subpopulations. The group of full deficiencies (0.05%) can, however, be separated into two subgroups, a group with subtotal deficiency with MPXI between −50 and −60 and a group with total deficiency with values <−60.

19.2.2 Pathology of MPO Deficiency

Total MPO deficiency has been for a long time claimed to be without serious pathological consequences, except a certain predisposition to fungal infections [2, 3]. We have now detected 77 cases of total and subtotal deficiency, of which 68 could be followed. Figure 19.3 summarizes the different pathologies encountered. Even if some individuals are perfectly healthy in spite of their defect, our figures represent a clear accumulation of inflammatory and infectious complications. It becomes even more evident when we compare the MPO-deficient population with a similar control population of 72 probands with normal MPO activity (Fig. 19.4) [4]. The difference in the sum of infectious problems for the two populations is highly significant. The χ^2-test with Yates' correction yields a value of 14.88, $\alpha < 0.01$. The frequency of life-threatening infections in the deficient group is slightly less significant: $\chi^2 = 4.66$, $0.025 < \alpha < 0.01$.

19.2.3 Protective Effect of MPO Deficiency Against Cardiovascular Diseases

The total absence of cases of myocardial infarction and coronam artheroma in the MPO-deficient group compared with the high frequency in the control group is striking and significant; the χ^2-test with Yates' correction yields a value of 6.08, $0.025 < \alpha < 0.01$. This suggests a protective effect of MPO deficiency against myocardial infarction and coronary atheroma. This could result from decreased lipoprotein oxidation due to the absence of hypochlorite normally generated by the peroxide-peroxidase reaction and scavenging of the antioxidant capacity. A larger number of cases are needed to clarify this question. Systematic determination of total antioxidant capacity could yield additional evidence. It would also be interesting to see to what extent similar protection is found in the much more numerous deficient heterozygotes.

We have not observed an increased incidence of malignancies, as reported by Lanza [5]. This high frequency might have been the result of bias by selection of pati-

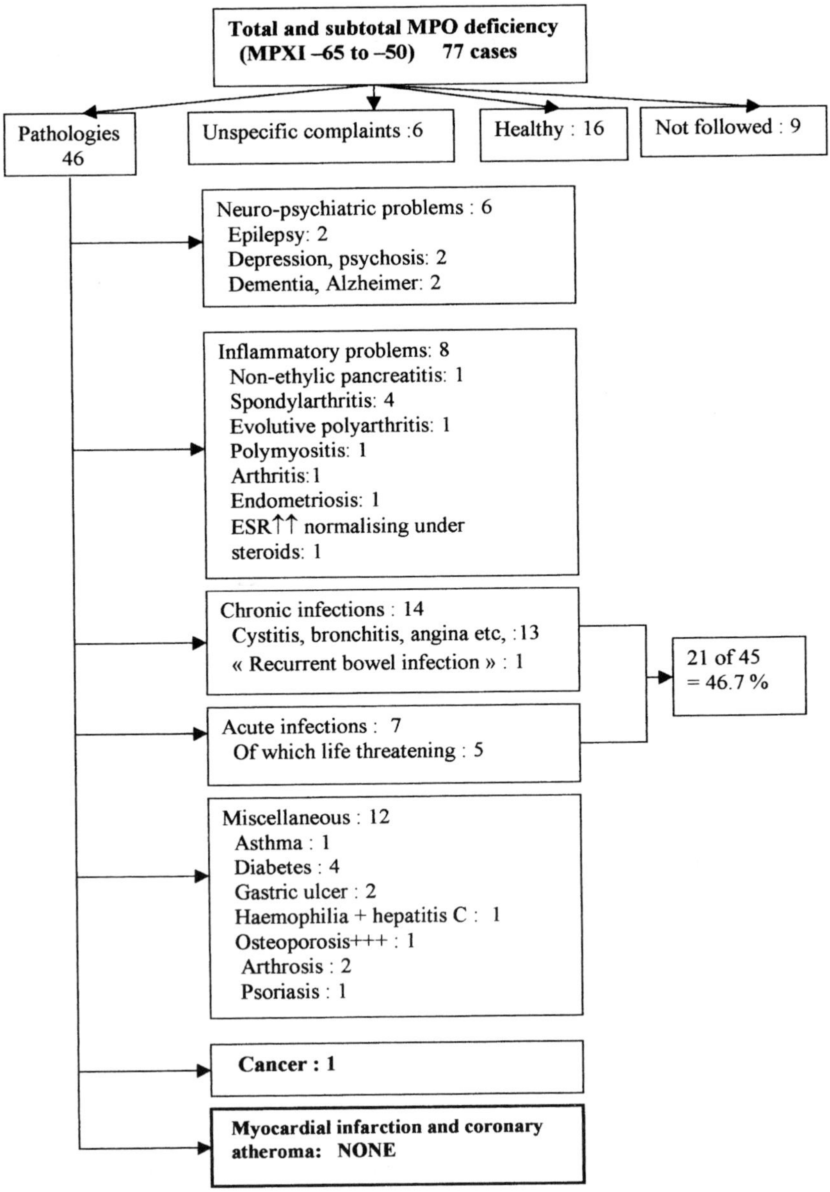

Fig. 19.3. Pathologies of 77 cases of total and subtotal MPO deficiency. Several pathologies may be shown by the same proband

ents from among those attending an oncology clinic, whereas our cases were from the general population.

We have not yet found a clear correlation between mild or partial MPO deficiency with specific pathologies.

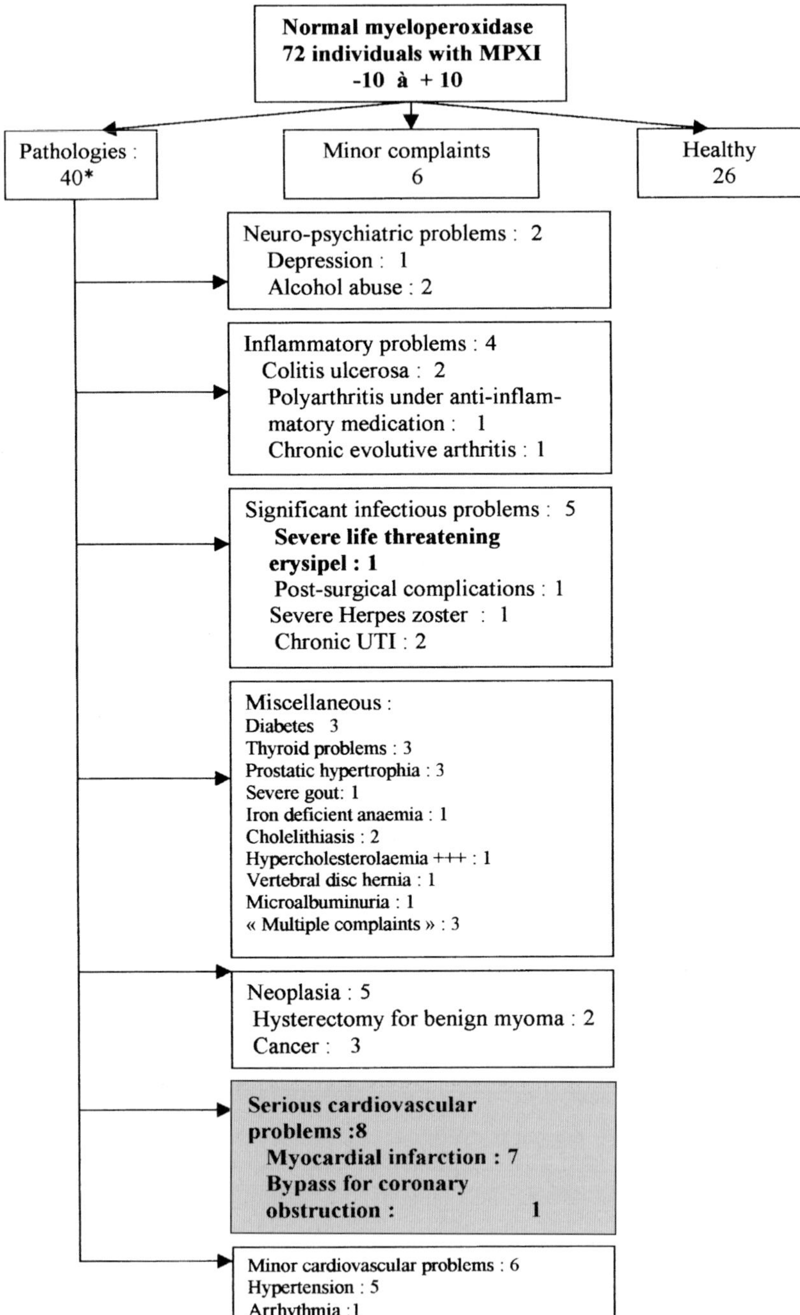

Fig. 19.4. Pathologies in a reference population of 72 individuals with normal MPXI selected at random from a village community. * Several individuals show more than one pathology

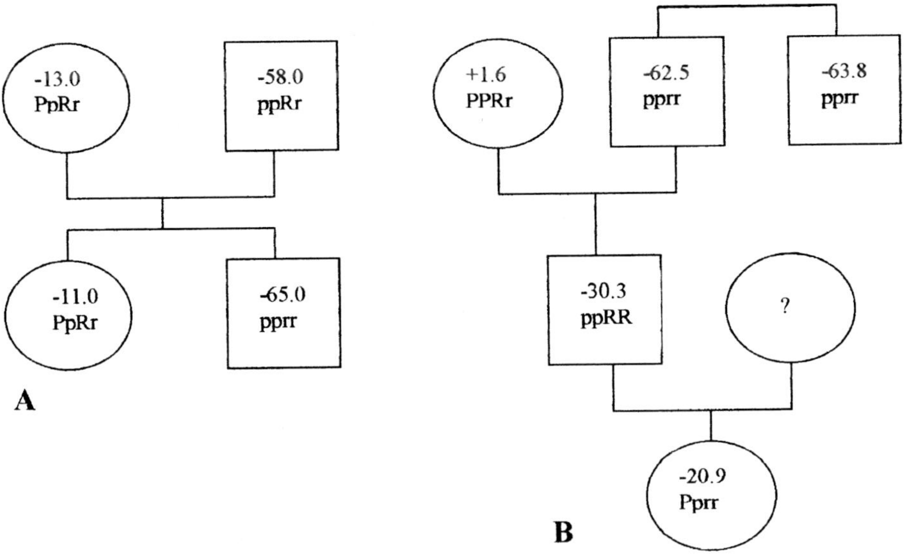

Fig. 19.5. Pedigrees of two families with MPO deficiency indicating the MPXI and the genotype derived from a bigenic model implicating a production gene P and a regulatory gene R

19.2.4
Genetics of MPO Deficiency

While screening for total MPO deficiency, we were able to perform a number of interesting family studies which have given some insight into the genetic transmission of this anomaly. MPO activity in the PMN of family members of totally deficient propositi vary within large limits and do not follow a pattern as would be expected for a monogenic, recessive disorder (PP, Pp, pp). Our findings agree much more with a bigenic model, involving two hypothetical genes, for instance a production gene P and a regulatory gene R. The combination of these two genes would yield nine genotypes and could explain the large range between normal (PPRR) and totally deficient (pprr) [6]. Figure 19.5 shows two hitherto unpublished pedigrees with hypothetical genotypes deduced from this scheme.

19.2.5
MPO Deficiency of Monocytes

We have reported an interesting observation of selective monocyte MPO deficiency in a highly anemic patient [7]. The leukocyte count was normal with high monocytosis. Monocytes appearing completely normal at microscopy were localized to the LUC area on the H-3 diagram (Fig. 19.6). The absence of MPO from all the monocytes was confirmed by peroxidase staining and considered indicative of their immaturity. Bone marrow examination led to the unexpected diagnosis of acute myelomonocytic leukaemia. Monocytes developed normal MPO at remission.

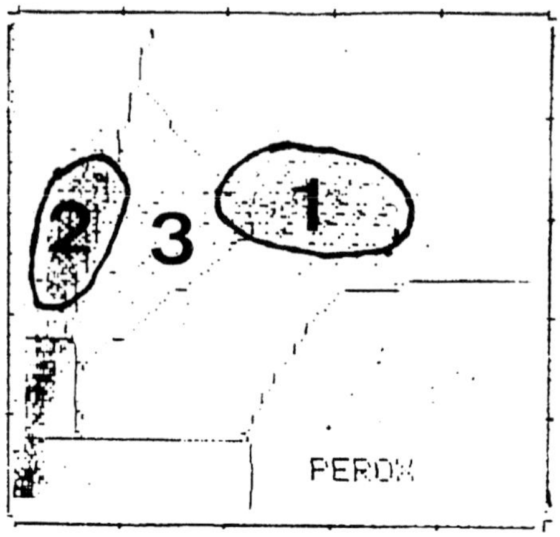

Fig. 19.6. MPO deficiency of monocytes

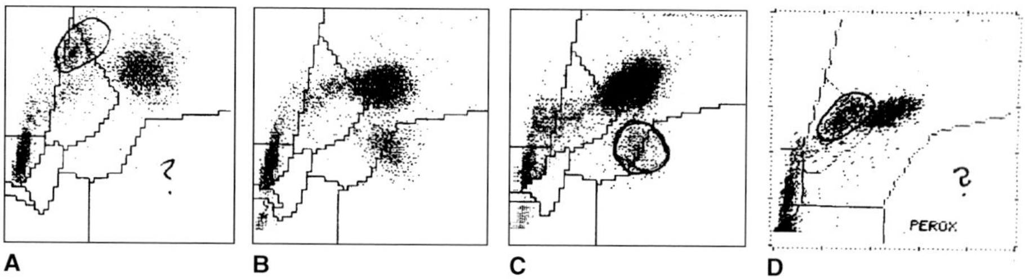

Fig. 19.7A-D. ADVIA displays. **A** Total human EPO deficiency. The eosinophil area is void, eosinophils appear as an additional cluster in the upper monocyte area. **B** Hypothetical partial human EPO deficiency. The cluster deviates towards the left indicating decreased EPO activity. **C** Normal dog. Important eosinophil cluster. **D** Dog with total EPO deficiency. The eosinophil cluster is void; eosinophils appear as an important additional cluster in the upper monocyte area (26% eosinophils on microscopy)

19.3 EPO Deficiency

Eosinophil peroxidase (EPO) deficiency is evident on the BAYER diagrams. The eosinophil cluster is absent from its usual location (Fig. 19.2A) and replaced by an abnormal cluster on top of the monocyte area (Fig. 19.7A). Peroxidase staining shows the eosinophils to be totally EPO-deficient. We have up to now observed 16 cases with this anomaly. Details are summarized in Table 19.1. In spite of the two cases of cancer, we see no clear correlation between EPO deficiency and definite pathologies. The presence of four probands of Mediterranean origin corresponds to the ethnic composition of the population of the Grand Duchy of Luxembourg, resulting from massive Portuguese and Italian immigration.

Family studies have shown completely normal location of the eosinophil cluster in the diagrams of all available parents and offspring [8]. In spite of this, we consider our deficient probands homozygous and the examined parents and offspring heterozygous and at least mildly deficient. This could be equivalent to the mild MPO deficiencies. We have, however, observed several routine cases presenting eosinophil clusters

Table 19.1. Details of 16 cases of EPO deficiency

Case no.	Sex	Age (years)	Ethnic origin	Eosinophils ($\times 10^3/\mu l$)	Anamnesis	Consulting for
1	M	82	Caucasian, Luxembourg	0.70	Operated for rectal adenocarcinoma in 1990	Minor complaints
2	M		Caucasian, Italy	0.66	No striking details	Minor complaints
3	F	55	Caucasian, Italy	0.52	No striking details	Minor complaints
4	F	82	Caucasian, Luxembourg	0.64	Mastectomy for adenocarcinoma in 1996	Routine check-up
5	F	77	Caucasian, Luxembourg	0.57	No striking details	Minor complaints
6	M	28	Caucasian, Luxembourg	0.17	No striking details	Complaining of abnormal fatigue
7	F	72	Caucasian, Luxembourg	0.25	Evolutive chronic polyarthritis	Therapy
8	F	88	Caucasian, Luxembourg	0.17	No striking details	Routine check-up
9	M	62	Caucasian, Luxembourg	0.24	Alcohol abuse	Weaning control
10	M	46	Caucasian, Germany	0.12	No striking details	Minor complaints
11	F	5	Caucasian, Luxembourg	0.56	No striking details	Minor complaints
12	F	37	Caucasian, Luxembourg	0.12	Allergic problems without confirmation by RAST	Minor complaints
13	F	37	Caucasian, Portuguese	0.34	Hypotensive episodes	Minor complaints
14	F	39	Pakistan	0.46	No striking details	Fatigue
15	F	43	Caucasian, Luxembourg	0.42	Allergic problems without confirmation by RAST	Fatigue
16	M	35	Caucasian, Portuguese	0.45	No striking details	Joint problems resulting from work as woodcutter

in an irregular position, suggesting decreased EPO activity (Fig. 19.7B). An anomaly of this type was observed in both father and son, suggesting its genetic character. Further family studies are necessary to verify the existence of more deficient heterozygotes among parents and offspring of totally EPO-deficient persons.

EPO is also found in dog eosinophils, which occupy a location close to that of the human cells in the PEROX diagrams (Fig. 19.7C). We have observed a clear case of canine total EPO deficiency, as shown in Fig. 19.7D. The area normally occupied by the eosinophils is void and replaced by an irregular, very dense cluster in the upper monocyte area. Pappenheim and peroxidase staining revealed 26% of peroxidase-free eosinophils! Unfortunately the "propositus" turned out to be a sterilized female mongrel dog of unknown origin, preventing the breeding of EPO-deficient dogs.

19.4 Summary

While no clear pathology has been found for total EPO deficiency, it is no longer possible to consider total MPO deficiency a simple genetic curiosity. Even if a potent immunological defense seems to compensate for the defect in a certain percentage of deficient individuals, an accumulation of inflammatory and infectious complications becomes evident, especially in comparison with a control group of arbitrarily selected probands with normal MPO activity. The idea of a protective effect of MPO deficiency against cardiovascular complications is tempting and calls for future investigations. Flow cytometry as performed by the BAYER hematological analyzers give an insight into peroxidase activity of neutrophils, eosinophils and monocytes, enabling the routine laboratory to select the cases and the families with MPO and EPO deficiencies from large numbers of patients and to direct these to more sophisticated molecular investigations to clarify the genetic transmission.

Acknowledgements. We express our gratitude to our colleagues Dr. C. Missenard, Dr. J. C. Nord, Dr. J. P. Hendriks and Dr. M. Michels for signalling cases of MPO deficiency and to the numerous medical colleagues who provided us with data concerning these cases.

References

1. Kutter D, Al Haidari K, Thoma J (1994) Myeloperoxidase deficiency: simple methods for its diagnosis and significance of different forms. Klin Lab 40: 342–346
2. Lehrer RI, Cline MJ (1969) Leukocyte myeloperoxidase deficiency and disseminated candidiasis. J Clin Invest 48: 1478–1488
3. Nauseef WM (1988) Myeloperoxidase deficiency. Hematol Oncol Clin North Am 2: 135–158
4. Kutter D (1998) Pathologie de la déficience subtotale et totale en myéloperoxidase. Bull Soc Luxemb Biol Clin 19: 75–80
5. Lanza F (1998) Clinical manifestation of myeloperoxidase deficiency. J Mol Med 76: 676–681
6. Kutter D, Niedner A, Lieners C (1994) Coexistence de deux populations distinctes de granulocytes neutrophiles, l'une totalement, l'autre partiellement déficiente en myéloperoxidase (in French). Bull Soc Sci Med Lux 131: 31–34
7. Kutter D, Dicato M, Ries F, Mahassen A, Delcourt A (1998) Deficience totale et sélective en myéloperoxidase des monocytes périphériques menant au diagnostic d'une leucémie nyélo-monocytaire aigüe. Bull Soc Sci Med Lux 134: 5–9
8. Kutter D, Verstraeten L (1998) Dépistage du déficit en peroxidase des granulocytes éosinophiles. Ann Biol Clin (Paris) 56: 205–208

Chapter 20

Prevalence of Inherited Myeloperoxidase Deficiency in Japan

Kazuo Suzuki, Hiroyuki Nunoi, Makoto Miyazaki, Fumikazu Koi

20.1 Introduction

Neutrophils possess potently active biological systems able to kill in vitro tumors as well as a wide range of microorganisms, including bacteria, fungi [1–3], and viruses [4–6]. To this end, neutrophils utilize directly microbicidal granule proteins as well as a variety of reactive oxygen intermediates. However, the most efficient oxygen-dependent system exploits the combined effects of hydrogen peroxide, generated by activation of the NADPH-dependent oxidase, and the granule protein myeloperoxidase (MPO; E.C. 1.11.1.7).

The broad range of microbicidal and tumoricidal activity of the MPO-hydrogen peroxide-halide system has been extensively documented [1]. Based on the importance of this system, one would predict that subjects whose neutrophils lack MPO would suffer severe and frequent infections, in effect paralleling the serious infectious morbidity associated with defective oxidase activity as seen in chronic granulomatous disease. However, there are few case reports of serious infectious complications associated with MPO deficiency [7]. The recent application of automated flow cytometry to quantitate leukocyte differential counts from peripheral blood has demonstrated that MPO deficiency is relatively common in the general population. Studies from Europe and from the United States have demonstrated a prevalence of complete deficiency of 1 in 2000 to 1 in 4000. We undertook these studies to determine the prevalence in the general population in Japan and to assess the impact of MPO deficiency on the prevalence or severity of infection in affected individuals.

20.2 Materials and Methods

20.2.1 Questionnaire

We distributed a questionnaire concerning the prevalence and classification of MPO deficiency to 167 clinical laboratories in Japan which use the Technicon H series of analyzers (H-1, H-1E, H-2, or H-3). The degree of MPO deficiency was classified into four types according to the MPO index (MPxi) determined by flow cytometry (Fig. 20.1) and done previously [7]. Of the laboratories contacted, 50 returned completed questionnaires.

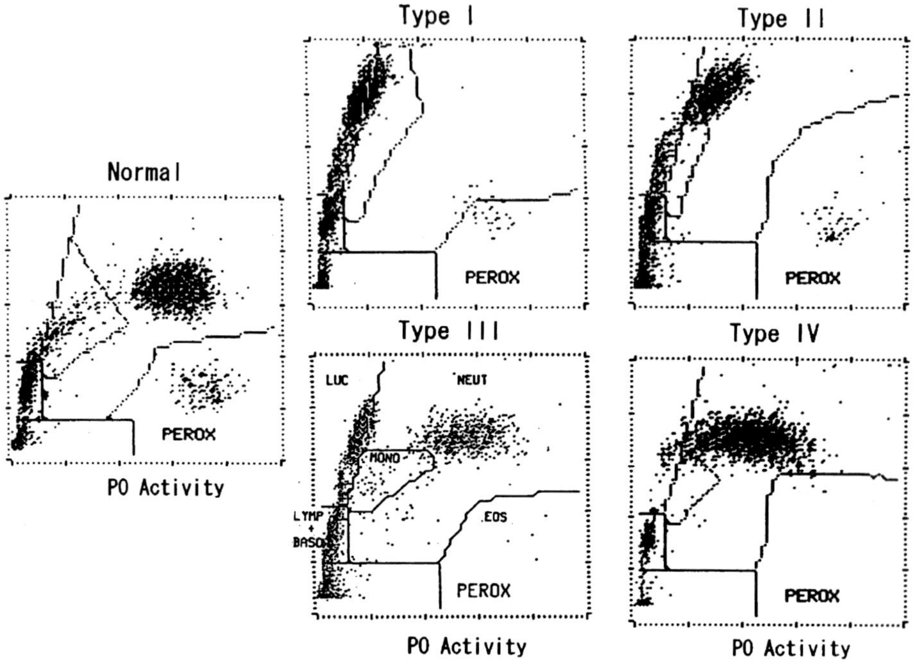

Fig. 20.1. Classification of abnormal profile of neutrophil counts in the cytogram. The cytograms were determined using the Technicon H series of analyzers. The procedure for displaying the cytogram is described in Materials and Methods

20.2.2 Analysis of MPO Deficiency

Venous blood from individuals was anticoagulated with EDTA and differential leukocyte counts were determined by automated flow cytometric analysis of peroxidase-stained samples using the Technicon H analyzer (Bayer-Sankyo Co., Tokyo). In brief, cells were fixed in formalin and peroxidase activity determined colorimetrically, using 4-chloro-1-naphthol as substrate. Four types of MPO deficiency were identified based on the MPxi (Table 20.1, Fig. 20.1) [8]. We defined complete MPO deficiency as an MPxi below −55 and partial deficiency as an MPIX between −55 and −35 [7].

Table 20.1. Classification of neutrophil abnormality determined with the Technicon H series of analyzers (peroxidase activity: normal, −10 to +10; theoretically completely negative, below −65, but below −50 absolute activity not determined, therefore completely negative below −50

Type	Activity	MPXI	Level of abnormal peroxide activity on the cytogram
I	Completely negative	<−50	Almost all neutrophils negative
II	Low activity	−50 to −35	Fewer than half the neutrophils negative
III	Multi-type		Two groups in peroxidase activity
IV	Diverse	Normal	Wide distribution

20.3 Results and Discussion

From the 50 laboratories surveyed, we identified 155 subjects with some degree of MPO deficiency, 26 of whom had complete deficiency and 129 partial deficiency (Table 20.2). Based on the populations studied, we estimate the prevalence of complete MPO deficiency to be 1 in 57,135 and partial deficiency to be 1 in 17,501 in the Japanese general population (Table 20.2). In similar studies in other countries, the prevalence of MPO deficiency has been reported to be much higher than we found in Japan. In Italy the prevalence is 1 in 4000 [9], in the United States 1 in 2000–4000 [10], and in Luxembourg 1 in 1801 [7]. The striking difference in prevalence between Japan and Western Europe and the United States most likely reflects the relative heterogeneity of the genetic background of the populations in the West. Precise definition of the genetic bases of the MPO deficiencies within each population may clarify the nature of the observed differences (Chapter 23).

Type	Cases	Prevalence
I	26	1/57,135
II	129	1/17,501
III	1	1/150,000 to 300,000
IV	None	

Table 20.2. Prevalence of MPO deficiency (average for each type)

A single gene on chromosome 17q21.3-q23 encodes MPO [11]. After cotranslational glycosylation, the 90-kDa apoproMPO acquires heme and undergoes posttranslational proteolytic processing to generate the lysosomal form of the enzyme, with subunits of 59 kDa and 14 kDa [12–20]. Several studies have assigned to MPO deficiency an autosomal recessive pattern of inheritance, but the pattern appears to be more complex. Indeed, Kutter et al. [7] have proposed a bigenic pattern of inheritance (see Chap. 19), reflecting alterations in regulatory as well as structural genes, although there are no data to support the involvement of regulatory elements. Three different genotypes underlying MPO deficiency have been described, all involving the MPO gene. Several MPO-deficient individuals have been described with a missense mutation whereby an arginine is replaced by a tryptophan at codon 569 (R569W) [21, 22] and Y173C [23]. In the presence of this mutation, MPO biosynthesis is limited to synthesis of the enzymatically inactive apoproMPO form; heme is not inserted and mature subunits are not formed [20, 22]. Romano et al. [24] have described two other mutations in the MPO gene in subjects with inherited MPO deficiency. In one case a missense mutation at codon 251 results in substitution of threonine for methionine whereas the other mutation reflects the deletion of 14 nucleotides from exon 9. The impact of these mutations on MPO biosynthesis is unknown. In addition, we are characterizing in the DNA of a Japanese subject with MPO deficiency, a novel mutation in a region of the MPO gene encoding the heavy subunit (Ohashi and Suzuki, unpublished results). Given these findings and the report of Tobler et al. [13] describing a pretranslational abnormality causing MPO deficiency, it is clear that a wide range of genotypes underlie inherited MPO deficiency.

The clinical sequelae of MPO deficiency remain somewhat subtle. Two subjects in our study had bacterial pneumonia, but the remaining individuals were healthy. In addition, in the Luxembourg study two subjects were found with complete MPO deficiency who had severe bacterial infections. The report of Lanza [25] suggests that tumors are more common in MPO-deficient subjects than in the population at large. None of the partially deficient subjects in our study had infections. Taken together, the data suggest that complete deficiency is a risk factor for severe infection [7] but the presence of even a low level of MPO activity represents an effective defense mechanism against infection. The recent availability of a murine model of MPO deficiency [26] provides an experimental approach to address questions about the contribution of MPO-dependent processes to overall host defense and to inflammatory events in general.

20.4 Summary

The microbicidal activity of the MPO-hydrogen peroxide-halide system has been implicated as the most efficient, oxygen-dependent antimicrobial element of neutrophil host defenses. Unexpectedly, individuals with MPO deficiency suffer few clinical consequences. In order to understand better the clinical impact of MPO deficiency, we surveyed several clinical hematology laboratories in Japan to assess the prevalence of MPO deficiency in the general population. MPO activity was determined by flow cytochemistry using the Technicon H series of automated systems. We identified 26 cases of complete MPO deficiency (prevalence 1 in 57,135), and 129 cases of partial deficiency (prevalence 1 in 17,501). The distribution of complete and partial deficiency differed among the laboratories studied.

Acknowledgements. We thank the many doctors and technicians in the 50 laboratories in Japan for precise analysis of patients with MPO deficiency and for the effort spent completing the questionnaire. We acknowledge as well Dr. William M. Nauseef, University of Iowa, for his valuable suggestions and Dr. Shuji Hashimoto, School of Health Sciences and Nursing, University of Tokyo, for his advice in statistical analysis.

References

1. Klebanoff SJ (1990) Myeloperoxidase – occurrence and biological function. In: Everse J, Grisham M (eds) Peroxidase in chemistry and biology. CRC Press, Boca Raton, 180–192
2. Klebanoff SJ (1980) Oxygen metabolism and the toxic properties of phagocytes. Ann Intern Med 93: 480–489
3. Clark RA, Klebanoff SJ, Einstein AB, Fefe A (1975) Peroxidase H_2O_2-halide system: cytotoxic effect on mammalian tumor cells. Blood 45: 161
4. Yamamoto K, Suzuki K, Suzuki K, Mizuno S (1989) Phagocytosis and ingestion of influenza virus by human polymorphonuclear leukocytes in vitro: electron microscopy studies. J Med Microbiol 28: 191–198
5. Yamamoto K, Miyoshi-Koshio T, Utuki Y, Mizuno S, Suzuki K (1991) Virucidal activity and viral protein modification by myeloperoxidase: a candidate for defense factor of human polymorphonuclear leukocytes against influenza infection. J Infect Dis 164: 8–14
6. Klebanoff SJ, Coobes RW (1992) Viricidal effect of polymorphonuclear leukocytes on human immunodeficiency virus-1: role of the myeloperoxidase system. J Clin Invest 89: 2014–2017
7. Kutter D, Haidari AI, Thoma J (1994) Myeloperoxidase deficiency: simple methods for its diagnosis and significance of different forms. Klin Lab 40: 342–346

8. Ornstein L, Ansley H, Saunders A (1976) Improving manual differential white cell counts with cytochemistry. Blood Cells 2: 557–585
9. Patriarca P, Cramer R, Tedesco F, Kakinuma K (1975) Studies on the mechanism of metabolic stimulation in polymorphonuclear leukocytes during phagocytosis. II. Presence of the NADPH oxidizing activity in a myeloperoxidase-deficient subject. Biochim Biophys Acta 385: 387–393
10. Nauseef WM, Root RK, Malech HL (1983) Biochemical and immunologic analysis of hereditary myeloperoxidase deficiency. J Clin Invest 71: 1297–1307
11. Inazawa J, Inoue K, Nishigaki H, Tsuda S, Taniwaki M, Misawa S, Abe T (1989) Assignment of the human myeloperoxidase gene (MPO) to bands q21.3-q23 of chromosome 17. Cytogenet Cell Genet 50: 135–136
12. Nauseef WM (1986) Myeloperoxidase biosynthesis by a human promyelocytic leukemia cell line: insight into myeloperoxidase deficiency. Blood 67: 865–872
13. Tobler A, Selsted ME, Miller CW, Johnson K.R, Novotny MI, Rovera G, Koeffler HP (1989) Evidence for a pretranslational defect in hereditary and acquired myeloperoxidase deficiency. Blood 73: 1980–1986
14. Andrew PC, Krinsky NI (1981) The reductive cleavage of myeloperoxidase in half, producing enzymatically active hemi-myeloperoxidase. J Biol Chem 256: 4211–4218
15. Anderson MR, Atkin CL, Eye HJ (1982) Intact form of myeloperoxidase from normal human neutrophils. Arch Biochem Biophys 214: 273–283
16. Olsen RL, Little C (1984) Studies on the subunits of human myeloperoxidase. Biochem J 220: 701–709
17. Chang KS, Trujillo JM, Cook RG, Stass SA (1986) Human myeloperoxidase gene: molecular cloning and expression in leukemia cells. Blood 68: 1411–1414
18. Suzuki K, Yamaha M, Abash K, Fujikura T (1986) Similarity of kinetics in three types of myeloperoxidase of human leukocytes and four types of HL-60 cells. Arch Biochem Biophys 245: 167–173
19. Homma T, Suzuki K, Kudo Y, Inagawa M, Mizuno S, Yamaguchi K, Tagawa M (1989) Preparation and characterization of monoclonal antibodies against human myeloperoxidase. Arch Biochem Biophys 273: 189–196
20. Nauseef WM, Cogley M, McCormick S (1996) Effect of the R569W missense mutation on the biosynthesis of myeloperoxidase. J Biol Chem 271: 9546–9549
21. Nauseef WM, Brigham S, Cogley M (1994) Hereditary myeloperoxidase deficiency due to a missense mutation of arginine 569 to tryptophan. J Biol Chem 269: 1212–1216
22. Nauseef WM, Cogley M, Bock S, Petrides PE (1998) Pattern of inheritance in hereditary myeloperoxidase deficiency associated with R569W missense mutation. J Leukoc Biol 63: 264–269
23. DeLeo FR, Goedken M, McCormick SJ, Nauseef WM (1998) A novel form of hereditary myeloperoxidase deficiency linked to endoplasmic reticulum/proteasome degradation. J Clin Invest 101: 2900–2909
24. Romano M, Dri P, Dadal L, Patriarca P (1997) Biochemical and molecular characterization of hereditary myeloperoxidase deficiency. Blood 90: 4126–4134
25. Lanza F (1998) Clinical manifestations of myeloperoxidase deficiency. J Mol Med 76: 676–681
26. Aratani Y, Koyama H, Nyui S-I, Suzuki K, Kura F, Maeda N (1999) Severe impairment in early host defense against *Candida albicans* in mice deficient in myeloperoxidase. Infect Immun 67: 1828–1836

Acquired and Inherited Forms of Myeloperoxidase Deficiency: Clinical and Hematological Features

Francesco Lanza, Angela Latorraca, Sabrina Moretti, Barbara Castagnari, Luisa Ferrari, Gianluigi Castoldi

21.1 Introduction

Myeloperoxidase (MPO), an iron-containing heme protein localized in the azurophilic granules of neutrophil granulocytes and in the lysosomes of monocytes, contributes to a potent bactericidal system which is effective against fungi, viruses, and tumor cells [1–5]. Unlike B and T lymphocytes, mature PMNs and monocytes lack clonal antigen-specific receptor molecules. However, as cellular components of the innate immune system, they possess a variety of surface proteins that serve as pattern recognition structures able to distinguish invading microbes or altered components of self. MPO and its precursor, proMPO, are expressed very early in myeloid differentiation and can be found in over 30% of $CD34^+$ bone marrow progenitor cells which coexpress CD45RA and possess in vitro clonogenic potential. MPO deficiency, as well as other congenital or acquired abnormalities of PMNs, have provided insights into the mechanisms by which certain cellular components mediate normal neutrophil functions, and have helped to delineate the effective role played by these systems in host defense against bacterial and fungal infections.

21.2 Classification of MPO Deficiency

MPO deficiency is now easy to diagnose using automated flow cytochemistry, and the application of this technology has provided a better understanding of this disorder. MPO deficiency can be divided into two subgroups: (1) primary or congenital and (2) secondary or acquired. The main clinical and hematological features of the two types of deficiency are summarized in Table 21.1.

Primary MPO deficiency has a genetic origin, occurs (in most cases) with different degrees of severity in more than one family member, and involves both the neutrophil and monocyte lineages [5–15]. Eosinophils and their marrow precursors are never involved in MPO deficiency, since the gene encoding eosinophil peroxidase is distinct from that for MPO.

Secondary MPO deficiency occurs in a wide variety of clinical situations, including hematological disorders (especially those involving the maturation processes of the granulocytic series), iron deficiency, infectious diseases, lead intoxication, diabetes mellitus, neuronal lipofuscinosis, renal failure, cardiovascular diseases, thrombotic diseases, and disseminated cancers in which it seems to be related to cytostatic drug administration (Table 21.2). MPO deficiency in these cases can be transient, and successful treatment of the underlying disease usually corrects the enzymatic defect

Table 21.1. Comparative analysis of clinical and hematological characteristics of primary vs secondary MPO deficiency

Clinical and biological features of MPO deficiency	Primary	Secondary
Heritable MPO activity in blood cells	Yes (small activity in partial MPO deficit)	No (variable from cell to cell)
MPO activity in bone marrow myeloid cells	Absent	Variable from cell to cell
Variation of MPO activity throughout life	Absent (detected in a few patients who develop additional MPO gene mutations)	Present (correction of the deficit after successful treatment of the underlying disease)
Peroxidase activity in eosinophils	Present (absent in Presentey anomaly)	Present (in a few cases decreased)
Type of gene mutation	Germline	Somatic
Concomitant diseases	Occasionally	Always

Table 21.2. Clinical situations which may be associated with MPO deficiency

Hematological neoplasms
Acute myeloid leukemia
Chronic myeloproliferative disorders (chronic myeloid leukemia, polycythemia vera, myelofibrosis with myeloid metaplasia)
Myelodysplastic syndromes
Disseminated cancers
Drugs (cytotoxic agents, antiinflammatory compounds)
Iron deficiency
Lead intoxication
Cardiovascular disease
Thrombotic diseases (endothelial cell damage)
Renal transplantation
Severe infectious diseases
Diabetes mellitus
Neuronal lipofuscinosis
Pregnancy

(Table 21.1) [3, 10–12, 16–18]. In the large majority of patients with the acquired form, the MPO deficiency has been found to be partial and limited to a fraction of the neutrophil population. Affected subjects have circulating cells with normal MPO activity together with cells having a mild to moderate deficiency in MPO expression [16]. MPO-deficient PMNs obtained from patients with myelodysplastic syndromes (MDS) generally present dysplastic changes, i.e. nuclear abnormalities, cytoplasmic hypogranulation, reduced expression of several membrane and intracellular antigens, and functional anomalies [16, 17]. However, based on the report of Becker and Pflüger [17], authors do not suggest distinguishing between acquired and inherited deficiencies solely on the basis of the analysis of granular enzymes (elastase-like protease, lactoferrin, and MPO), since molecular and genetic methods should be preferred. There is also evidence that subjects with severe iron deficiency may show a significant reduction in MPO activity, which may be corrected by iron treatment [19].

However, it should be pointed out that some controversy exists in the literature regarding the causative role of several pathological conditions in the pathogenesis of secondary MPO defect. In some of these cases, family studies are of great value in defining the origin of the enzymatic defect [20–22].

Studies concerning the immunoreactivity of MPO-deficient PMNs have shown that the levels of MPO antigens, analyzed by means of anti-MPO monoclonal antibodies, seem to decline concurrently with enzyme activity [10, 15, 27]. These findings suggest that in these patients, the lack of detectable activity in neutrophil granulocytes reflects the absence of MPO-related protein rather than the presence of a dysfunctional form of MPO. In contrast, the near normal antigenic reactivity of neutrophils from a large proportion of patients with acquired MPO deficiency (e.g. MDS) indicates the presence of a functionally inactive form of the enzyme in these patients [16, 27].

21.3 Clinical Aspects

Although MPO is involved in the killing of certain microorganisms, no particular susceptibility to persistent or severe infections has been noted in the large majority of MPO-deficient patients reported thus far in the literature. Sporadically, severe infections in MPO-deficient patients have been documented but the frequency of such cases is very low, less than 5% of reported MPO-deficient subjects (Table 21.2) [5–12]. Visceral infections, especially with species of *Candida*, have been observed predominantly in MPO-deficient patients who also have diabetes mellitus. Nevertheless, clinical data and the demonstration that the microbicidal and fungicidal activity of MPO-deficient neutrophils are only slightly impaired, indicate that there are MPO-independent mechanism(s) for microbicidal activity, which are particularly efficient in subjects with total MPO deficiency [3, 6, 8, 14]. Some reports have shown that MPO-deficient PMNs respond to stimuli with a prolonged respiratory burst and an increase in hydrogen peroxide production, factors which may compensate for their absence of peroxidase [3, 6, 8]. These findings have not been confirmed in other studies dealing with individuals affected by congenital MPO deficiency [5, 9, 14]. The high level of alkaline phosphatase activity of MPO-deficient neutrophils as well as the increased number of blood eosinophils have been associated with congenital MPO deficiency, thus supporting the concept that alternative mechanisms could be particularly efficient in PMNs lacking the MPO-H_2O_2-halide system [14]

Since most MPO-deficient subjects do not suffer from infections, maintenance prophylactic therapy with antibiotics is not recommended. Caution is required in patients with diabetes mellitus, who suffer from a high tendency toward localized and systemic infections, and concomitant complete MPO deficiency. In these cases prompt and prolonged therapy with antibiotics is usually necessary to control infections, and aggressive use of parenteral antibiotics is crucial during infection with resistant microorganisms or species of fungi.

Still of great controversy is the relationship between congenital MPO deficiency and the occurrence of malignancies. In a previous study, it has been found that subjects with total MPO deficiency show a high incidence of malignant tumors. In a study by Lanza et al., 7 out of 12 individuals with complete MPO deficiency had solid or hematological tumors [14]. Kitahara et al. [6], and Parry et al. [7] have also observed a consistent association of cancers with MPO deficiency (Table 21.3).

Table 21.3. Reported incidence of infectious disorders and malignancies in congenital MPO-deficient subjects

Author	Patients (n)	Infections	Tumors
Kitahara, 1981	32	2 (streptococcal cellulitis, aseptic meningitis)	3 (testicular, lymphoma)
Parry, 1981	28	4 (two disseminated infections: candidiasis, bacteremia)	1 (craniopharyngioma)
Cramer, 1982	45	No clinical data	No clinical data
Larrocha, 1982	12	No clinical data	None
Nauseef, 1983	5	No clinical data	No clinical data
Lippi, 1983	72	3	2 (Hodgkin lymphoma, myeloproliferative disorder)
Lanza, 1984	10	None	2
Lanza, 1987	58	1 (urinary tract)	7 (breast, uterus, myeloma)
Kutter, 1994	23	2 (sepsis caused by *Staphylococci*)	None
Becker, 1994	106	6	16
Muller, 1996	1	1	/
Nguyen, 1997	1	1 (disseminated candidal dermatitis)	/
London, 1997	Case control study	/	Reduced lung cancer risk
Petrides (unpublished)	3	Osteomyelitis	1 (breast cancer)
Kutter, 1998	52	12 (serious infectious or inflammatory problems)	1
Lanza, 1998	28	2	14 (10-year follow-up)

More recently, we have confirmed in a series of 28 patients our previous observation that there is a strong association between total MPO deficiency and malignant disease. Over a follow-up period of 10 years, 14 out of 28 individuals with total MPO deficiency had concomitant malignant disease. The basis for this association is still uncertain, but previous investigations on the cytotoxic activity of total MPO-deficient neutrophils, showing depressed lytic action against human tumor cells, could explain this observation [23–25]. The involvement of neutrophil MPO in mediating tumor cell cytotoxicity has been demonstrated by the inhibition of cytotoxicity by azide and cyanide, by the markedly impaired activity of MPO-deficient neutrophils, and by the specific correction of this defect on addition of purified MPO [31].

Based on these data, it can be speculated that the neutrophil MPO system plays a key role in the tumor surveillance of the host, thus suggesting a causative role for MPO in the pathogenesis of cancer in patients affected by complete MPO deficiency. However, these results have not been confirmed by subsequent studies. London et al. have observed a reduced lung cancer risk in MPO-deficient subjects [26]. Since MPO activates benzo-a-pyrene and aromatic amine in tobacco smoke and generates carcinogenic free radicals in the lung, individuals with a reduction in neutrophil MPO activity may be at decreased risk for lung cancer. Large population studies with long-term follow-up will more clearly define the role of neutrophil MPO in tumor surveillance.

21.4
Diagnostic Value of Anti-MPO Monoclonal Antibodies in Leukemia

Another area of interest pertinent to MPO expression is related to the immunophenotype analysis of leukemic cells [28–32]. Recent advances in the use of monoclonal antibodies (mAbs) and flow cytometry have greatly facilitated the definition of the maturation stages of both lymphoid and myeloid hemopoietic lineages. In addition, many authors have used this approach to classify acute leukemias. The combined use of anti-MPO and anti-lysozyme mAbs is of clinical and experimental importance in several areas of leukemia research. These tools are important for the definition of FAB M0 subtype, the identification of biphenotypic leukemias as well as their distinction from acute leukemia with minimal phenotypic deviation, and the recognition of the so-called "pure" M5 FAB subvariety.

Expression of MPO has been found in the large majority of acute myeloid leukemia (AML) patients. However, the levels of expression vary significantly among the various FAB subtypes. FAB M1-M3 subtypes are characterized by intermediate-bright expression for MPO, whereas the M5 FAB subvariety has little or no expression of MPO. MPO is also expressed in about 40–50% of AML with minimal evidence of myeloid differentiation (M0), thus indicating that MPO-negative blasts by cytochemical assay may be stained by anti-MPO mAbs. These data suggest that anti-MPO antibodies may recognize the enzymatically inactive precursor forms of MPO [30]. However, there is evidence that the degree of positivity for MPO protein is strictly dependent on the ability of the mAbs to recognize the precursor forms of MPO. It is likely that a comparative analysis of different MPO mAbs could be of great value for the lineage assignment of acute leukemic blasts of unknown origin, and for the classification of the different types of MPO deficiency, especially in those individuals who have a varying degree of MPO activity in PMNs and monocytes. It is likely that the testing of a wide panel of MPO mAbs towards recombinant constructs of MPO precursor with or without the pro-MPO region could improve the definition of maturation stages of myelomonocytic hemopoietic lineages, redelineate the classification of AML, and of acquired and inherited forms of MPO deficiency.

21.5
Summary

MPO deficiency is the most common inherited disorder of neutrophils. Apart from the inherited form, acquired MPO deficiency has been described, although both the incidence and the clinical relevance of this anomaly are still uncertain. Despite the primary role of the MPO system in the killing of microorganisms and foreign cells, subjects with hereditary total or partial MPO deficiency have no apparent increase in infection rates. Infectious diseases, especially with species of *Candida*, have been observed predominantly in MPO-deficient patients who also have diabetes mellitus, but the frequency of such cases is very low – less than 5% of reported MPO-deficient subjects. Recent data have indicated that individuals with primary total MPO deficiency may show a high incidence of malignant tumors. The basis for this association is still uncertain, but previous investigations on the cytotoxic activity of MPO-deficient neutrophils, showing depressed lytic action against human tumor cells, could explain this observation. It is likely that the analysis of a larger number of MPO-deficient subjects and a sharper focus on their family members may allow a better understanding of this disorder, and, hopefully, provide insights into the clinical implication related to this neutrophil dysfunction.

References

1. Nauseef WM, Olsson I, Arnljots K (1988) Biosynthesis and processing of myeloperoxidase. A marker for myeloid cell differentiation. Eur J Haematol 40: 97–110
2. Nauseef W (1986) Myeloperoxidase biosynthesis by a human promyelocytic leukemia cell line: insight into myeloperoxidase deficiency. Blood 67: 865–872
3. Stendahl O, Coble BI, Dahlgren C, Hed J, Molin L (1984) Myeloperoxidase modulates the phagocytic activity of polymorphonuclear neutrophil leukocytes. Studies with cells from a myeloperoxidase-deficient patient. J Clin Invest 73: 366–373
4. Clark RA, Szot S (1981) The myeloperoxidase-hydrogen peroxide-halide system as effector of neutrophil mediated tumor cell cytotoxicity. J Immunol 126: 1295–1301
5. Cech P, Papathanassiou A, Boreux G, Roth P, Miescher PA (1979) Hereditary myeloperoxidase deficiency. Blood 53: 403–411
6. Kitahara M, Eyre HJ, Simonian Y, Atkin CL, Hasstedt SJ (1981) Hereditary myeloperoxidase deficiency. Blood 57: 888–893
7. Parry MF, Root R K, Metcalf JA, Delaney KK, Kaplow LS, Richar WJ (1981) Myeloperoxidase deficiency. Ann Intern Med 95: 293–301
8. Cramer R, Soranzo MR, Rottini GD, Bramezza M, Cirielli S, Patriarca P (1982) Incidence of myeloperoxidase deficiency in an area of northern Italy: histochemical, biochemical and functional studies. Br J Haematol 51: 81–87
9. Larrocha C, De Castro MF, Fontan G, Viloria A, Chacon JLF, Jimenez C (1982) Hereditary myeloperoxidase deficiency: study of 12 cases. Scand J Haematol 29: 389–397
10. Nauseef WM, Root RK, Malech HL (1983) Biochemical and immunologic analysis of hereditary myeloperoxidase deficiency. J Clin Invest 71: 1297–1307
11. Cappelletti P, Lippi U (1983) Hereditary myeloperoxidase deficiency: a rare condition? Diagnostic possibilities of a differential white cell autoanalyzer (Hemalog-D). Hematologica 68: 736–741
12. Kaplow LS (1986) The relationship of myeloperoxidase activity to neutrophil maturity and other hematologic indicators of infection. Blood Cells 12: 153–165
13. Cech P, Schneider P, Bachmann F (1982) Partial myeloperoxidase deficiency. Acta Haematol 67: 180–184
14. Lanza F, Fietta A, Spisani S, Castoldi GL, Traniello S (1987) Does a relationship exist between neutrophil myeloperoxidase deficiency and the occurrence of neoplasms? J Clin Lab Immunol 22: 175–180
15. Bos AJ, Weening RS, Hamers MN, Wever R, Behrendt H, Roos D (1982) Characterization of hereditary partial myeloperoxidase deficiency. J Lab Clin Med 99: 589–600
16. Moretti S, Lanza F, Spisani S, Latorraca A, Rigolin GM, Giuliani AL, Castoldi GL, Traniello S (1994) Neutrophils from patients with myelodysplastic syndromes: relationship between impairment of granular contents, complement receptors, functional activities and disease status. Leuk Lymphoma 13: 471–477
17. Becker R, Pflüger KH (1994) Myeloperoxidase deficiency: an epidemiological study and flow cytometric detection of other granular enzymes in myeloperoxidase deficient subjects. Ann Haematol 69: 199–203
18. D'Onofrio G, Mancini R, Vallone R, Alfano G, Candido A, Palla M, Mango G (1983) Acquired neutrophil myeloperoxidase deficiency: an indicator of subclinical activation of blood coagulation? Blood Cells 9: 455–464
19. Murakawa H, Bland C, Willis W, Dallman PR (1987) Iron deficiency and neutrophil function: different rates of correction of the depressions in oxidative burst and myeloperoxidase activity after iron treatment. Blood 69: 1464–1468
20. Bozdech MJ, Bainton D, Mustacchi P (1980) Partial peroxidase deficiency in neutrophils and eosinophils associated with neurologic disease. Am J Clin Pathol 73: 409–416
21. Lanza F, Castoldi GL, Masotti M (1988) Eosinophil peroxidase deficiency detected by the Technicon H1 system. Ann Haematol 56: 143–145
22. Kutter D, Mueller-Hagedorn S, Forges T, Glaesener R (1995) A case of eosinophil peroxidase deficiency. Ann Haematol 71: 315–317

23. Lanza F, Giuliani AL, Amelotti F, Spisani S, Traniello S, Castoldi GL (1988) Depressed neutrophil mediated tumor cell cytotoxicity in subjects affected by hereditary myeloperoxidase deficiency and secondary neoplasia. Hematologica 73: 355–358
24. Lanza F (1998) Clinical manifestation of myeloperoxidase deficiency. J Mol Med 76: 676–681
25. Clark R, Szot S (1981) The myeloperoxidase-hydrogen peroxide-halide system as effector of neutrophil-mediated tumor cell cytotoxicity. J Immunol 126: 1295–1301
26. London SJ, Lehman TA, Taylor JA (1997) Myeloperoxidase genetic polymorphism and lung cancer risk. Cancer Res 57: 5001–5003
27. Lanza F, Latorraca A, Musto P, Ferrari L, Moretti S, Zabucchi G, Carotenuto M, Castoldi GL (1991) Cytochemically unreactive neutrophils from subjects with myeloperoxidase (MPO) deficiency show a complex pattern of immunoreactivity with anti-MPO monoclonal antibodies: a flow cytometric and immunocytochemical study. Ann Haematol 63: 94–100
28. Knapp W, Strobl H, Majdic O (1994) Flow cytometric analysis of cell-surface and intracellular antigens in leukemia diagnosis. Cytometry 18: 187–198
29. Lanza F, Moretti S, Latorraca A, Ferrari L, Castagnari B, Castoldi GL (1996) Cytofluorimetric measurement of cytoplasmic myeloperoxidase by permeabilisation of blood cells with the FACS lysing solution. Cytometry 24: 292–295
30. Van Der Shoot CE, Daams GM, Pinkster J, Vet R, Von Dem Borne A (1990) Monoclonal antibodies against myeloperoxidase are valuable immunological reagents for the diagnosis of acute myeloid leukaemia. Br J Haematol 74: 173–178
31. Lanza F, Latorraca A, Moretti S, Castagnari B, Ferrari L, Castoldi GL (1997) Comparative analysis of different permeabilisation methods for the flow cytometric measurement of cytoplasmic myeloperoxidase and lysozyme in normal and leukemic cells. Cytometry 30: 134–144
32. Benè M, Castoldi GL, Knapp W, Ludwig WD, Matutes E, Orfao A, Veer MB van't, on behalf of the European Group for the Immunological Characterization of Leukemias – EGIL (1995) Proposals for the immunological classification of acute leukemias. Leukemia 9: 1783–1786

Myeloperoxidase-Positive and Negative Granulocytes: Investigations by Flow Cytometry

Gernot Bruchelt, Claudia E. Gerber, Peter Bader, Matthias Zipfel, Thomas Klingebiel, Rupert Handgretinger, Dietrich Niethammer

22.1 Introduction

Myeloperoxidase (MPO) is one of the key enzymes in the oxygen-dependent microbicidal activity of neutrophils and monocytes. Different forms of MPO deficiency have been described, caused by disturbances in RNA or protein at the genomic level of development [1]. Despite the fact that the prevalence of complete MPO deficiency is rather high (about 1:4000), and despite the importance of MPO in the generation of HOCl and some other compounds with high antimicrobial activity [2], a surprisingly low number of severe complications have been reported in MPO-deficient subjects. The introduction of hematological routine analyzers, which differentiate white blood cells according to their peroxidase activity, in combination with FACS-based analyzers has proved to be helpful for the diagnosis and to yield new insights into the genetic origin of MPO deficiency. A summary of the present state of the art of biochemistry and clinical aspects of MPO is given in a recently published series [3].

MPO and MPO-dependent intracellular processes can be estimated by different flow cytometric methods. Using different anti-MPO monoclonal antibodies and FACS analysis, the presence of MPO within permeabilized cells can be determined as a method of distinguishing between MPO-positive and -negative leukemias, of subclassifying myeloid leukemias, and of identifying biphenotypic leukemias [4]. Similar approaches can be used to examine agonist-dependent activation of the oxidative burst. For example, one method loads neutrophils with the nonfluorescent dichlorofluorescein diacetate, which can penetrate cell membranes. Cytosolic proteases clip the ester group thereby leaving the substrate trapped in the cell. Activated neutrophils generate H_2O_2 that converts the substrate to a fluorescent dye, a process that is strongly enhanced by the presence of peroxidase [5]. A third method that will be described in more detail in this chapter is the determination of peroxidase activity in leukocytes by using the Bayer-Technicon analyzers. These hematological systems are widely used in blood cell analysis and allow the determination of granulocytes and monocytes negative for MPO and eosinophil peroxidase (EPO) [6]. In addition to its use for routine analysis, the system can be used for measuring in whole blood functional processes associated with the MPO activity. Some examples of the application of the Bayer H3 for different experimental questions will be demonstrated in the following sections.

22.2
Principle and Advantages of MPO-Activity Measurement in Neutrophils of Whole Blood Samples

Figure 22.1a shows the distribution of white blood cells measured by the Bayer H3. The cell cluster location towards the right side of the plot ("PEROX" channel, *x*-axis) correlates with the increasing peroxidase-activity of the cells (monocytes < neutrophils < eosinophils). The mean MPO activity in neutrophils of a single subject is expressed as mean peroxidase index (MPXI). After analysis of several thousand healthy donors, the normal MPXI range was arbitrarily set between −10 and +10, including more than 90% of a healthy population [6].

Figure 22.1a shows a typical plot of granulocytes in whole blood samples compared to that of isolated granulocytes (Fig. 22.1b). In the "BASO" channel that is simultaneously plotted with the PEROX channel, only the area of polymorphonuclear cells (neutrophils and eosinophils) is still present after the gradient step isolation procedure using Histopaque 1119/1077, demonstrating a high purity of the granulocyte fraction which is confirmed by microscopic analysis. However, in the majority of granulocyte preparations, there is a significant shift of neutrophils to the left (*x*-axis, PEROX channel). This consequent decrease in MPXI is probably caused by partial degranulation of MPO. Additionally, a distinct portion of neutrophils is shifted to the

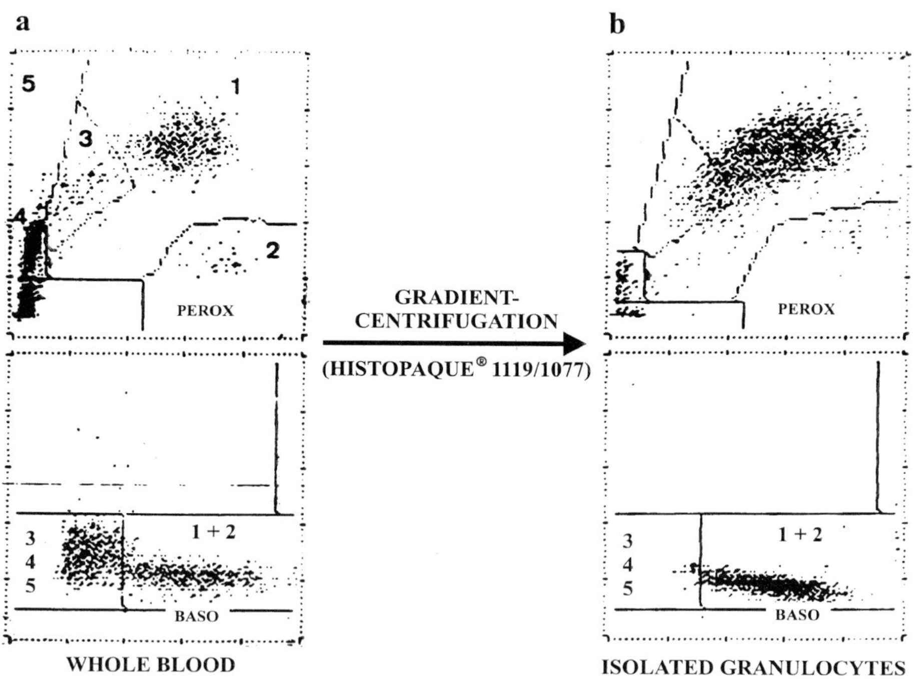

Fig. 22.1a,b. H3-plot of isolated granulocytes (**b**) from whole blood (**a**). After isolation, the neutrophil cluster (*1*) is broadened and somewhat shifted to the left after isolation (decrease in MPXI, PEROX plot). Only a few mononuclear cells [monocytes (*3*) and lymphocytes (*4*), located in the left rectangle of the BASO plot] are still present after isolation (*2* eosinophil granulocytes)

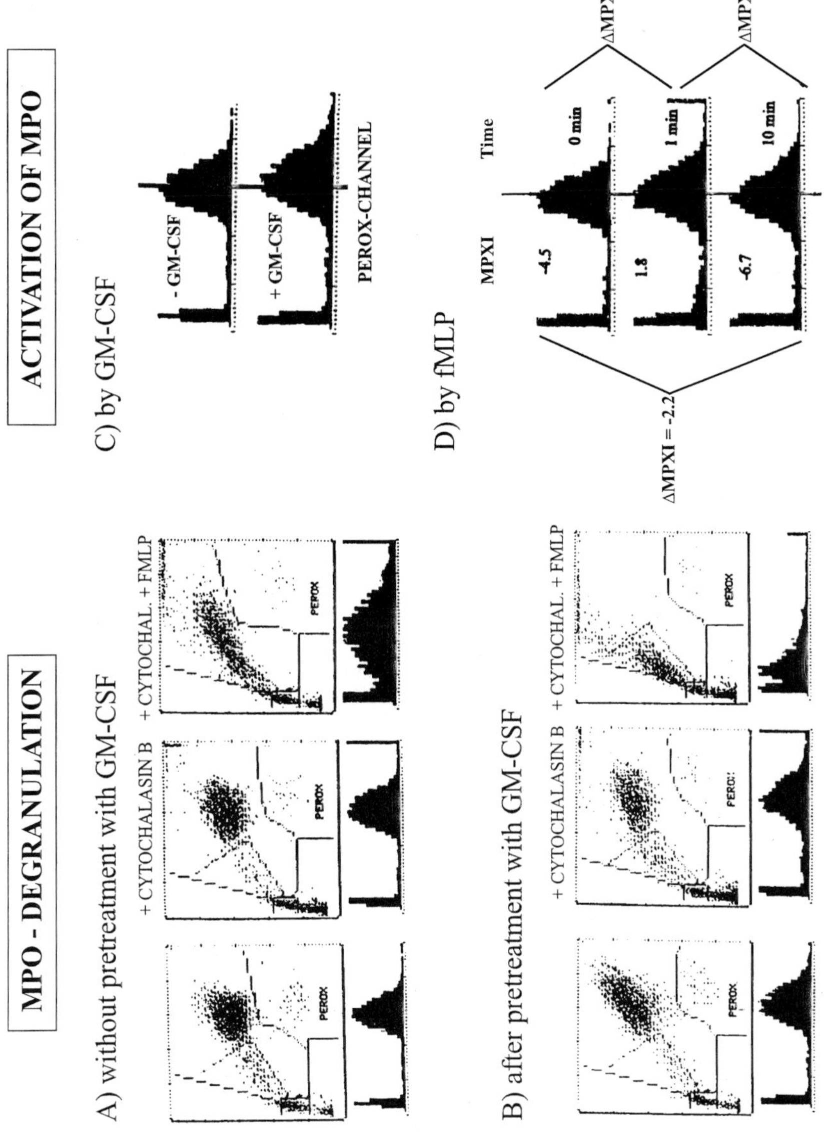

Fig. 22.2. A,B Release of MPO from neutrophils. **A** Whole blood samples were treated with cytochalasin B (5 µg/l) for 10 min and thereafter with 10^{-7} *M* fMLP for 1 min. **B** Same procedure as before, but whole blood samples were pretreated with 500 U/l GM-CSF. **C** Whole blood samples before and after treatment with 500 U/l GM-CSF as single agent for 2 h. **D** Shift of neutrophils after treatment with 10^{-7} *M* fMLP as a single agent for 1 min (increase in MPXI) and 10 min (drop in MPXI)

right, indicating that the MPO in these cells may be activated during the isolation procedure. Due to the high fragility of isolated granulocytes, it is advantageous that some determinations of MPO-dependent processes can be carried out in whole blood samples immediately after blood drawing. We therefore sought to determine whether it is possible to measure processes associated with changes in the amount and/or activity of MPO in whole blood after manipulation with different stimuli. Theoretically, one would predict that a shift to the left (reduction in MPXI) will occur when MPO is released from granulocytes or when MPO activity is inhibited, and that, in contrast, if the MPO activity could be enhanced, a shift to the right (increase in MPXI) should result. Experiments performed in order to test these predictions are shown in Fig. 22.2.

22.3 Degranulation of Neutrophils Measured by Determination of Remaining MPO Within the Cells Before and After Stimulation (Fig. 22.2)

Whole blood was incubated with *N*-formyl-methionyl-leucyl-phenylalanine (fMLP). In order to enhance the degranulation process, the samples were pretreated with cytochalasin B. As predicted, an intense shift of the neutrophils to the left was observed immediately after addition of fMLP (Fig. 22.2A). These effects were even more pronounced when the blood samples were preincubated with GM-CSF, a cytokine well known for its ability to prime degranulation [7] (Fig. 22.2B). The loss of MPO from activated granulocytes was confirmed by measuring MPO in the supernatant by an EIA [8].

22.4 Treatment of Whole Blood with GM-CSF and fMLP (Fig. 22.2)

GM-CSF does not only prime degranulation but also other agonist-dependent processes associated with granulocyte activation [9]. Treatment of whole blood with GM-CSF, applied as a single reagent, led to a significant increase in MPXI ($+4.75 \pm 2.2$, $n = 7$) compared to incidental fluctuations (ΔMPXI 0.83 ± 0.36, $n = 9$), indicating a transformation of MPO to a more activated state inside the granulocytes (Fig. 22.2C). Whether this process represents a direct activation, a removal of an endogenous inhibitor, a shift in pH, or the involvement of other factors is presently not known. Even more surprising was the observation that within 1 min of treatment with fMLP in the absence of cytochalasin B an increase in intracellular MPO activity occurred (ΔMPXI $+ 6.66 \pm 2$, $n = 6$), although a small part of MPO is simultaneously released into the supernatant [8]. Only some minutes after start of the incubation, MPXI drops due to continual release of MPO (Fig. 22.2D)

22.5 MPO Deficiency After Bone Marrow Transplantation: a Case Report

The usefulness of a hematological analyzer that differentiates white blood cells on the basis of peroxidase activity is shown in the following case report of an unusual bone marrow transplantation (Fig. 22.3). A 14-year-old girl with Philadelphia chromosome-positive acute lymphoblastic leukemia (ALL) was referred for allogeneic bone marrow transplantation from her HLA-compatible sister. The immunological work-up of the donor and the recipient was performed in the referring hospital. Before transplantation, the calculation of the donor's blood count was prepared for the first time on a Bayer H3 at our hospital, revealing that she was completely MPO-deficient (MPXI −62). The donor's MPO deficiency had not been identified previously, since

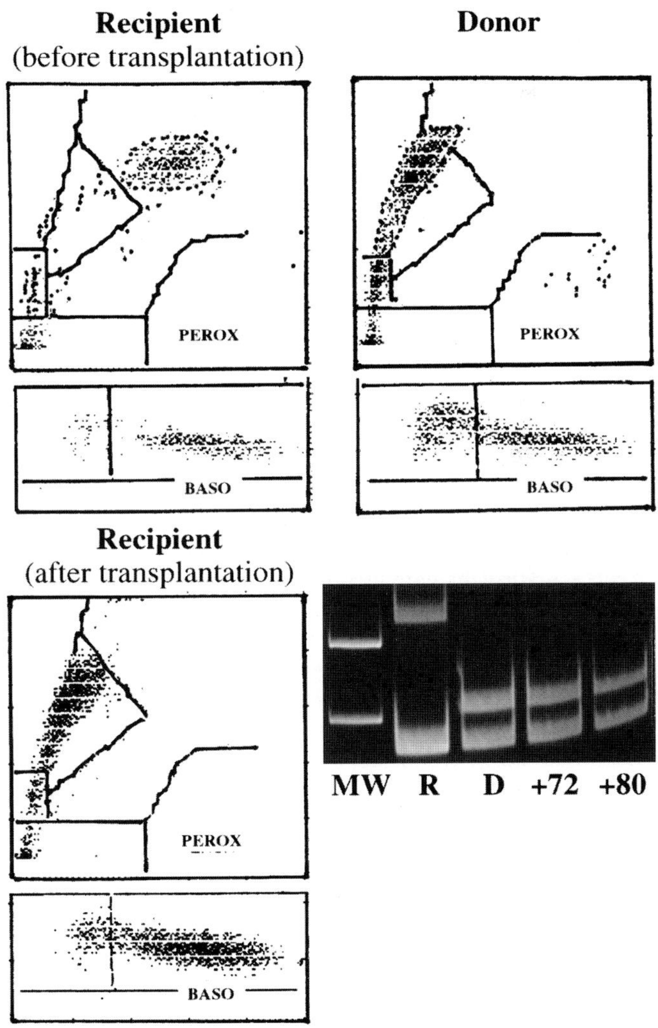

Fig. 22.3. H3 plots of whole blood samples of the recipient (MPXI−9) and the donor (MPXI−62) before bone marrow transplantation, and of the recipient after successful engraftment showing no MPO-positive neutrophils in the PEROX plot (day +80 after transplantation). The corresponding pattern of VNTR-PCR is also shown (*MW* molecular weight marker; *R* VNTR-PCR of the recipient before transplantation; *D* VNTR-PCR of the donor; *+72,+80* VNTR-PCR of the recipient on day 72 and 80, respectively, after bone marrow transplantation)

hematological studies performed up to this point were on cell counters which do not differentiate white blood cells on the basis of peroxidase activity. Consequently, the patient (recipient) who had only a slightly reduced MPO activity (MPXI −9) is now, after successful engraftment of donor marrow, completely MPO-deficient. Analysis of chimerism by PCR of a variable number of tandem repeat (VNTR) sequences showed a complete chimerism on days 72 and 80, indicating that all hematopoiesis was of donor origin. This was confirmed by the pattern of neutrophils in the PEROX diagram of the Bayer H3. In the special case of this transplantation, mixed chimerism would be easily detectable on the Bayer H3 by the occurrence of neutrophils with the original MPO activity of the recipient. An early detection of a mixed chimerism is important in order to start intervention since this situation is associated with a higher risk of rejection of the donor's graft or with a relapse of the underlying malignancy [10]. Of course, the knowledge of MPO deficiency requires specialized supervision with respect to bacterial and fungal infections of this immunologically weakened patient.

Additionally, this unique sequence of events posed another important question. Would this patient have a higher risk for experiencing a relapse or for developing a second malignancy due to a reduced immune surveillance because of her newly acquired MPO deficiency? Close monitoring of her hematologic picture and clinical status will address these important issues.

22.6 MPO Deficiency and Cancer Risk

There are some reports in the literature indicating that patients with MPO deficiency may have a higher risk for cancer [4, 11]. It has been suggested that one reason may be a reduced immune surveillance against cancer cells in MPO-deficient subjects, since MPO-deficient granulocytes show a reduced capacity to kill some tumor cells in vitro [12]. However, the granulocyte-mediated killing obviously depends strongly on the features of the target tumor cells. For example, we have found that granulocytes are nearly unable to kill neuroblastoma cells, even after activation with GM-CSF (Fig. 22.4a) [13]. However, when neuroblastoma cells are coated with antibodies directed against the ganglioside GD_2, which is present at high concentration on the cell membranes of most neuroblastoma cells, an effective killing can be obtained. Anti-GD_2 antibodies have been used quite successfully in the clinical treatment of neuroblastoma, and therefore, granulocyte-based antibody-dependent cellular cytotoxicity (ADCC) may contribute to the tumor cell killing also in vivo [14]. However, reactive oxygen intermediates do not play any role in this killing process. During the ADCC reaction, luminol-dependent chemiluminescence, a measure of peroxidase-dependent formation of reactive oxygen species (ROS), could not be detected [15]. In like fashion, killing of neuroblastoma cells is not reduced in the presence of inhibitors like azide, catalase or SOD [13]. It has even been observed that granulocytes of patients with NADPH-oxidase deficiency (chronic granulomatous disease) kill neuroblastoma cells much better than do normal granulocytes (Fig. 22.4b). Therefore, we conclude that the assumption that granulocytes that are not able to produce ROS have only a reduced immune surveillance capacity against malignant cells, including those generated by MPO-dependent processes, is not valid in this general sense.

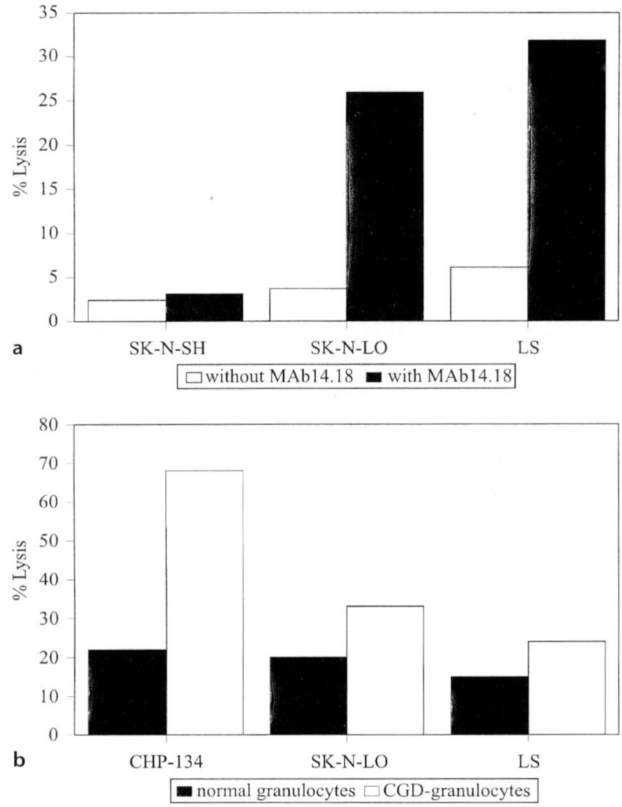

Fig. 22.4a,b. Destruction of human neuroblastoma cells by granulocytes. **a** Killing of the neuroblastoma cell lines SK-N-SH (lacking GD_2 expression), SK-N-LO and LS in the absence (*open bars*) and presence (*black bars*) of anti-GD_2-antibody mAb14.18 (^{51}Cr-release assay, effector to target ratio 60:1). **b** Killing of neuroblastoma cells SK-N-LO, LS and CHP-134 coated with mAb14.18 by normal granulocytes (*open bars*) and granulocytes from patients with chronic granulomatous disease (*black bars*) (^{51}Cr-release assay, effector to target ratio 40:1)

Nevertheless, the observation by some clinicians that MPO-deficient patients may have a higher risk for cancer is striking, but could reflect other factors than the proposed reduced immune surveillance. It is well known that ROS generated by activated granulocytes can cause mutations in the DNA of surrounding cells and therefore also cause malignant transformation [16, 17]. In particular, H_2O_2 is known as an excellent agent for indirect generation of DNA damage and mutations on adjacent cells, since it can easily penetrate cell membranes. MPO-deficient phagocytes could therefore be even more harmful than normal neutrophils, since they have an increased respiratory burst compared to normal monocytes and granulocytes [18]. Furthermore, in the absence of MPO, H_2O_2 cannot be converted to HOCl that is unable to penetrate cell membranes and to directly injure the DNA of neighboring cells [19]. Therefore, MPO-deficient granulocytes could cause more mutations, and this process may be another reason for the potentially higher cancer risk of MPO-deficient persons. However, it should be stressed that other investigators have not found evidence for a higher risk of MPO-deficient persons for getting cancer compared to the risk to normal subjects [6]. The situation is even more complex, if additional factors are considered. For example, it has been reported that MPO deficiency reduces the risk of smoking-induced lung cancer [20], since MPO-dependent processes have been found to enhance the carcinogenic potential of some of the compounds present in tobacco smoke. Additional large-population epidemiological studies are needed to clarify the link, if any, of MPO deficiency and altered risk for malignant disease.

22.7 Summary

The use of hematological routine analyzers that differentiate white blood cells by measuring peroxidase activity has substantially enlarged the knowledge of MPO-dependent processes under physiological and pathophysiological conditions. In addition to its advantages for performing routine blood cell counts, it allows the analysis of a number of MPO-dependent processes in whole blood without the need for prior isolation of the highly fragile and easily activated granulocytes.

Using the Bayer Technicon H3, different MPO-dependent processes have been analyzed in whole blood samples, in particular the influence of GM-CSF and fMLP on MPO content and/or activity of granulocytes. GM-CSF enhances intracellular MPO activity (increase in MPXI), but supports its release after incubation with fMLP. fMLP also increases MPO activity within the first minute after incubation, but MPXI drops after prolonged incubation due to a continuous release of MPO. Finally, the case of a bone marrow transplantation from a MPO-deficient donor to a nearly MPO-normal recipient suffering from ALL is described, and the consequences of MPO deficiency in regard to the risk of developing cancer are discussed.

References

1. Selsted ME, Miller CW, Novotny MJ, Morris WL, Koeffler HP (1993) Molecular analysis of myeloperoxidase deficiency shows heterogeneous patterns of the complete deficiency state manifested at the genomic, mRNA, and protein levels. Blood 82: 1317–1322
2. Weiss SJ (1989) Tissue destruction by neutrophils. New Engl J Med 320: 365–376
3. Petrides PE, Nauseef WM (1998) Molecular and clinical aspects of neutrophil peroxidase deficiency: multidisciplinary approaches on an international scale (editorial). J Mol Med 76: 659–698
4. Lanza F (1998) Clinical manifestation of myeloperoxidase deficiency. J Mol Med 76: 676–681
5. Bass DA, Parce JW, Dechatelet LR, Szdjda P, Seeds MC, Thomas M (1983) Flow cytometric studies of oxidative product formation by neutrophils: a graded response to membrane stimulation. J Immunol 130: 1910–1917
6. Kutter D (1998) Prevalence of myeloperoxidase deficiency: population studies using Bayer-Technicon automated hematology. J Mol Med 76: 669–675
7. Kauffman SE, DiPersio JF, Gasson JC (1989) Effects of human GM-CSF on neutrophil degranulation in vitro. Exp Hematol 17: 800–804
8. Zipfel M, Carmine Th, Gerber C, Niethammer D, Bruchelt G (1997) Evidence for the activation of myeloperoxidase by fMLP prior to its release from granulocytes. Biochem Biophys Res Commun 232: 209–212
9. DiPersio JF, Abboud CN (1992) Activation of neutrophils by granulocyte-macrophage colony-stimulating factor. In: Coffey RG (eds) Granulocyte responses to cytokines. Marcel Dekker, New York Basel Hong Kong, pp 457–484
10. Bader P, Beck J, Frey A, Schlegel PG, Hebarth H, Handgretinger R, Einsele H, Niemeyer C, Benda N, Faul C, Kanz L, Niethammer D, Klingebiel T (1998) Serial and quantitative analysis of mixed hematopoietic chimerism by PCR in patients with acute leukemias allows the prediction of relapse after allogeneic BMT. Bone Marrow Transplant 21: 487–495
11. Lanza F, Fietta A, Spisani S, Castoldi GL, Traniello S (1987) Does a relationship exist between neutrophil peroxidase deficiency and the occurrence of neoplasms? J Clin Lab Immunol 22: 175–180

12. Clark RA, Szot S (1981) The myeloperoxidase-hydrogen peroxide-halide system as effector of neutrophil mediated tumor cell cytotoxicity. J Immunol 126: 1295–1301
13. Bruchelt G, Handgretinger R, Firlbeck G, Kimmig A, Dopfer R, Reisfeld RA, Treuner J, Niethammer D (1989) Lysis of neuroblastoma cells by the ADCC-reaction: granulocytes of patients with chronic granulomatous disease are more effective than those of healthy donors. Immunol Lett 22: 217–220
14. Handgretinger R, Bader P, Dopfer R, Klingebiel T, Reuland P, Treuner J, Reisfeld RA, Niethammer D (1992) A phase I study of neuroblastoma with the anti-ganglioside GD2 antibody 14.G2a. Cancer Immunol Immunother 35: 199–204
15. Bruchelt G, Handgretinger R, Kimmig A, Goeke B, Siedner R, Reisfeld RA, Niethammer D, Treuner J (1989) Effects of granulocytes on human neuroblastoma cells measured by chemiluminescence and chromium-51 release assay. J Biolum Chemilum 3: 93–96
16. Weitzman SA, Gordon LI (1990) Inflammation and cancer: role of phagocyte-generated oxidants in carcinogenesis. Blood 76: 655–663
17. Dizdaroglu M, Olinski R, Doroshow JH, Akman StA (1993) Modification of DNA bases in chromatin of intact target human cells by activated human polymorphonuclear leukocytes. Cancer Res 53: 1269–1272
18. Locksley RM, Wilson CB, Klebanoff SJ (1983) Increased respiratory burst in myeloperoxidase-deficient monocytes. Blood 62: 902–909
19. Schraufstätter IU, Browne K, Harris A, Hyslop PA, Jackson JH, Quehenberger O, Cochrane ChG (1990) Mechanisms of hypochlorite injury of target cells. J Clin Invest 85: 554–562
20. London StJ, Lehman TA, Taylor JA (1997) Myeloperoxidase genetic polymorphism and lung cancer risk. Cancer Res 57: 5001–5003

Mutation Analysis for Genotype-Phenotype Relationships in Myeloperoxidase Deficiency

Petro E. Petrides, Susanne Bock, Chuanbing Zang

23.1 Introduction

My interest in myeloperoxidase (MPO) deficiency arose in the early 1990s when I began caring for a young female (RT) who had been in good health until the age of 11 years when she had developed osteomyelitis in her right tibia in 1979. Despite surgical intervention she had relapsed at the same site 2 years later. In 1988 she underwent partial resection of the right fourth rib because of the osteomyelitis. A specific microbiological etiology was not recovered from any of her operative specimens, blood or wound cultures. Antibiotic therapy was empiric and included agents with a broad range of activity. At no time were fungi isolated, serologic tests for *Candida* positive or empiric antifungal therapy used. Since her last surgical procedure she had been stable clinically requiring episodic courses of antibiotic treatment because of recurrent pain at the site of her previous infections. There was no evidence of diabetes mellitus.

23.2 Identification of MPO Deficiency in the Patient with Chronic Osteomyelitis

When I started to search for the underlying cause of the chronic osteomyelitis I also had a blood film stained for peroxidase activity which turned out to be negative (Fig. 23.1a). Blood films of the parents of RT when stained both were peroxidase-positive (Fig. 23.1b,c). Immunoblots of granulocytes isolated from RT probed with MPO antibodies showed the presence of the 90-kDa precursor protein but no subunits of mature MPO, indicating a failure in proteolytic processing (Fig. 23.2). Since the gene for human MPO had been cloned (reviewed in reference 2), the genomic DNA became amenable to genetic analysis. The MPO gene is a single gene of approximately 14 kb, composed of 11 introns and 12 exons, and localized on the long arm of chromosome 17 (Fig. 23.3).

23.3 Identification of the Mutation Involved in MPO Deficiency in This Patient

The first evidence for a genetic defect was the identification of aberrant restriction endonuclease digests of DNA when compared to normal individuals [3, 4]. In the family of RT as well as four unrelated families from the United States, a 2.6 vs 2.1 kb fragment was found when their DNA was digested with Bgl II and subjected to Southern blot analysis (Fig. 23.4). Further investigation [5, 6] determined that this RFLP

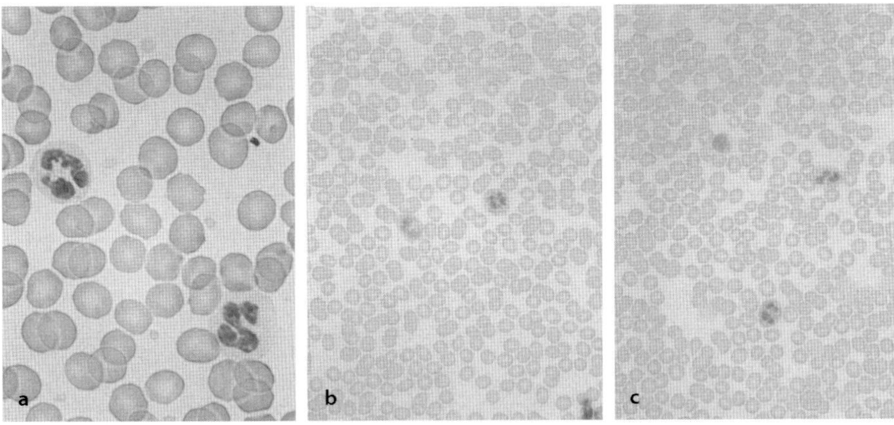

Fig. 23.1a–c. Peroxidase stain on blood films of patient RT (**a**), her mother RoT (**b**) and her father GT (**c**)

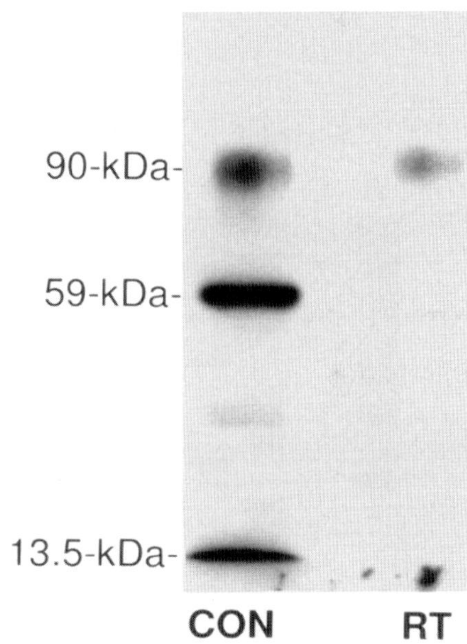

Fig. 23.2. Immunoblot of granulocytes from RT (from reference 1)

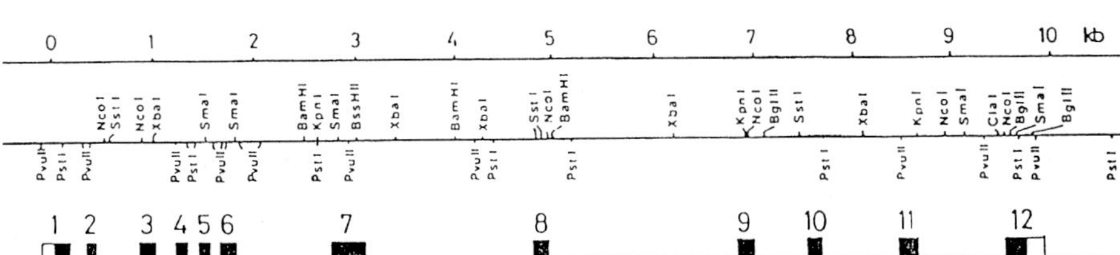

Fig. 23.3. Structure of the human MPO gene (from reference 2)

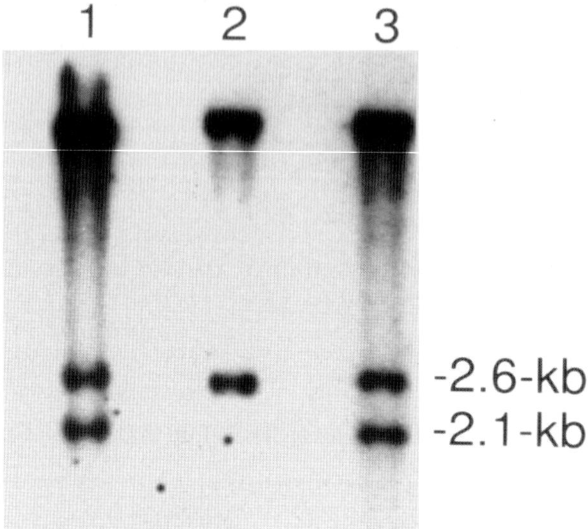

Fig. 23.4. RFLP-analysis of the DNA of RT (*1*), her father (*2*) and her mother (*3*) (from reference 1)

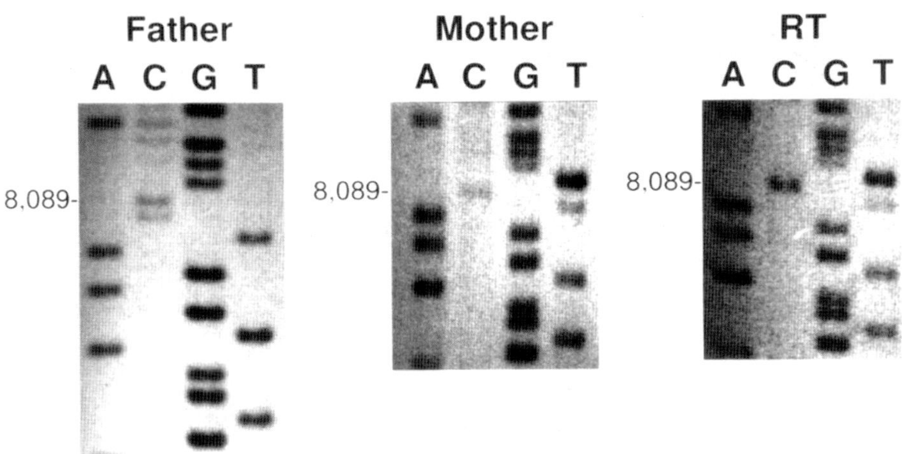

Fig. 23.5. Sequence analysis of exon 10 of the MPO gene of RT and her parents (from reference 1)

reflected a mutation in the MPO gene, thereby creating a recognition site for Bgl II (AGATCT) not present in normal individuals (AGATCC) (Fig. 23.5). This missense mutation caused by a C → T transition (8089C → T) results in substitution of tryptophan for an arginine at codon 569 (R569 W). This residue is conserved in eosinophil peroxidase (EPO) and lactoperoxidases (LPO) but replaced in thyroid peroxidase (TPO) by a threonine residue. The mutation results in a maturational arrest at the stage of apoproMPO. Thus the R569 W gene product is enzymatically inactive and is not processed or targeted to the mature lysosomal form of MPO.

Table 23.1. Mutant alleles in MPO deficiency (amino acid residue numbers counting from the N-terminal residue are shown in parentheses)

R569W (403)	R in MPO, EPO, LPO; T in TPO
Y173C (7)	Y in all peroxidases
M251T (85)	M in MPO and EPO; L in LPO and TPO
G501S (335)	G in all peroxidases

Similar R → W mutations occur in phenylketonuria (R408 W) and acute intermittent porphyria (R116W, R167W, R173W or R201W). In the latter case the PBG-deaminase protein is involved where the loss of the arginine side-chain leads to the interruption of stabilizing salt bridges within the protein or between the enzyme protein and its substrate. C → T transitions are mutational hotspots. Methylated cytosine residues can be deaminated and thus lead through a keto-enol tautomeric change to thymine. These residues pair with adenine residues during DNA replication, which causes the C → T transition. Meanwhile three more mutations in MPO deficiency have been identified (Table 23.1). In addition to substitutions that cause missense mutations, structural alterations in the regulatory parts of the MPO gene have been postulated to cause pretranslational defects (reviewed in reference 2)

23.4 Compound Heterozygosity in MPO Deficiency

RT has a complete loss of MPO enzymatic activity and immunoreactivity but is heterozygous for the R569 W mutation. She had inherited this mutation from her mother RoT who has – like her father – detectable peroxidase staining on blood films. However, when the MPO activity of her parents' neutrophils was quantitated and analyzed immunochemically, the semiquantitative nature of the peroxidase stain became obvious (Table 23.2). Since her father GT, who lacks the R569W mutation, had about 75%

Table 23.2. MPO activity, immunoreactivity and the presence of the R569 W mutation in the family of RT [1]

	MPO activity	Immunoreactivity	R569W
RT (daughter)	0	0	+/–
GT (father)	73	74	–/–
RoT (mother)	53	47	+/–

Table 23.3. Mutation analysis in various genetic disorders

Disease	Protein	Gene size (kb)	Number of mutations
Cystic fibrosis	CFTR	250	500
Phenylketonuria	Ph-Hydroxylase	90	240
Gaucher disease	β-glucocerebrosidase	7.3	80
Breast cancer	BCRA1	8	235
Acute porphyria	PBG-desaminase	10	130
PNH	GPI-synthase	17	100
MPO deficiency	MPO	14	4

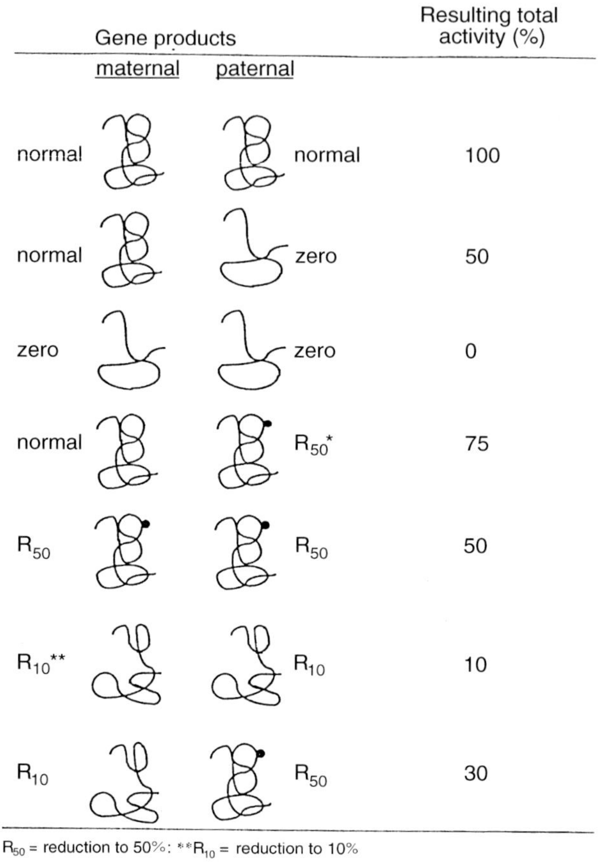

Fig. 23.6. Compound heterozygosity: combination of different alleles causes a different genotype and phenotype (e.g. resulting enzymatic activity) (from reference 2)

normal MPO enzymatic activity and immunoreactivity, he must have had an abnormal but still unknown MPO allele which he passed to his daughter. RT is a compound heterozygote for two mutant MPO alleles, as the combination of R569 W and the unknown but abnormal paternal allele causes a complete loss of activity.

Mutation analysis in other genetic diseases has revealed that more than 500 different mutations can occur in a given gene (Table 23.3). Some of these mutations occur in relatively high frequency while others are observed only in certain families and are therefore called private mutations. The presence of many mutations means that a large number of different combinations of mutated alleles determine the genotype. Figure 23.6 illustrates how the combination of different gene products with varied residual enzymatic activity can yield a large spectrum of enzymatically defined phenotypes. This variability explains in part the large clinical spectrum (phenotype) observed in various genetic disorders. In MPO deficiency the situation is further complicated since the enzyme is formed by proteolysis from a precursor and forms a dimer which has to be transported through the endoplasmic reticulum into the azurophilic granules.

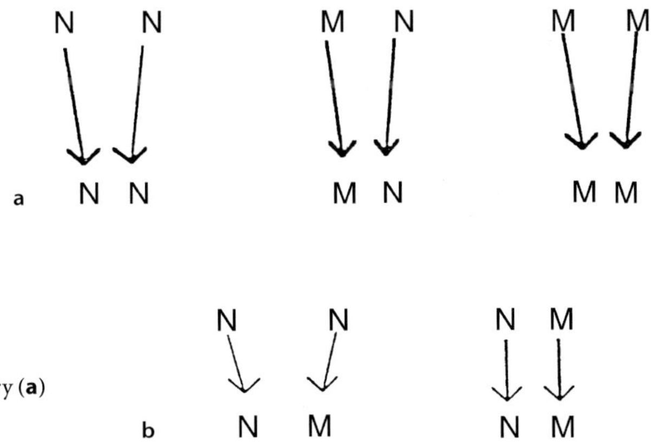

Fig. 23.7a,b. Genetics of hereditary (**a**) and acquired (**b**) MPO deficiency (from reference 2)

23.5 Hereditary and Acquired Forms of MPO Deficiency

MPO deficiency can be either acquired or inherited, depending upon whether germline and/or somatic mutations are involved. If germline mutations are involved both alleles have to be mutated to yield the clinical phenotype (Fig. 23.7a). Since hematopoietic stem cells are involved, MPO deficiency can also be acquired. For instance, a mutation in one allele could be present in the germline (without any clinical evidence); a second hit in a hematopoietic stem cell would then cause MPO deficiency. This would be analogous to what is observed in paroxysmal nocturnal hemoglobinuria. This leads to the presence of two coexisting granulocyte populations. Their quantitative ratio depends upon the growth tendency of each clone (Fig. 23.7b). This model would explain the observation by Kutter (Chap. 19) of two granulocyte populations in one individual.

23.6 Unanswered Questions in MPO Deficiency

With the observations reported here and in other chapters, several questions arise: Is MPO deficiency a disease or does it cause another disease? Does the type of mutation determine whether an individual is symptomatic? Do additional genes determine whether MPO deficiency in an individual has clinical consequences? Do clinical manifestations attributable to MPO deficiency arise at any age? Why is MPO deficiency so frequent (Table 23.4)? Is there a potential survival advantage for gene carriers? What could the value be of this potential survival advantage? Does a reduction of MPO activity cause less DNA or protein damage? What could be the disadvantage of MPO deficiency? Is it associated with increased susceptibility to viral, bacterial and fungal infections or to carcinogens? For the screening of a large number of mutations the PCR-DGGE approach has been proved valuable [7]. Only the determination of the genetic basis of this trait, the correlation of the genetic basis and clinical manifestations will provide answers to these questions.

Table 23.4. Frequency of homozygotes (compound heterozygotes) and gene carriers in autosomal recessive disorders

	Homozygotes	Gene carriers (or compound heterozygotes)
MPO deficiency:		
In Luxembourg	1:2,500	1:20
In Japan	1:50,000	?
Cystic fibrosis	1:2,500	1:20
Phenylketonuria	1:10,000	1:50
Gaucher disease	1:40,000	?

23.7 Summary

The MPO gene is a 14-kb gene located on the long arm of chromosome 17. Four mutations have so far been identified in individuals with MPO deficiency. As in other genetically determined diseases many more mutations will be identified, and some of them may predominate. This could be different in the various ethnic groups. In order to understand why some individuals with this inherited deficiency develop clinical symptoms (and become "patients") whereas others do not (and are "gene carriers"), mutation analyses of large numbers of patients are required to obtain precise genotype-phenotype relationships. Mutations can occur in germline and also in somatic cells. Most individuals will turn out to be compound heterozygotes displaying various amounts of residual MPO activity. Collection of clinical and molecular data through an international registry will allow for the establishing of genotype-phenotype relationships in regard to this genetic trait [8].

References

1. Nauseef W, Cogley M, Bock S, Petrides PE (1998) Pattern of inheritance in hereditary myeloperoxidase deficiency associated with the R569W missense mutation. J Leukoc Biol 63: 264–269
2. Petrides PE (1998) Molecular genetics of myeloperoxidase deficiency. J Mol Med 76: 688–698
3. Nauseef WM (1989) Aberrant restriction endonuclease digests of DNA from subjects with hereditary myeloperoxidase deficiency. Blood 73: 290–295
4. Petrides PE, Cogley M, Bock S, Nauseef WM (1994) Molekularpathologie eines familiär bedingten Myeloperoxidasemangels. Med Klin 89: S1,154
5. Nauseef WM, Brigham S, Cogley M (1994) Hereditary myeloperoxidase deficiency due to a missense mutation of arginine 569 to tryptophane. J Biol Chem 269: 1212–1216
6. Kusuki M, Miller CW, Selsted ME, Koeffler HP (1994) Myeloperoxidase (MPO) gene mutation in hereditary MPO deficiency. Blood 83: 1935–1940
7. Petrides PE (1998) Acute intermittent porphyria: mutation analysis and identification of gene carriers in a German kindred by PCR-DGGE analysis. Skin Pharmacol Appl Skin Physiol 11: 374–380
8. Petrides PE, Nauseef WM (1998) Molecular and clinical aspects of neutrophil peroxidase deficiency: multidisciplinary approaches on an international scale. J Mol Med 76: 659–660

Diagnostic Significance of Antineutrophil Cytoplasmic Antibodies (ANCA) in Systemic Vasculitides

Loïc Guillevin, Bernard Jarrousse

24.1 Introduction

Antineutrophil cytoplasmic antibodies (ANCA) are the immunological hallmark of several systemic vasculitides. Since the description of ANCA-associated vasculitides [1], ANCA have been demonstrated to be associated with small-sized vasculitides, such as Wegener's granulomatosis (WG) [2], Churg-Strauss syndrome (CSS) [3–5] and microscopic polyangiitis (MPA) [6]. Although ANCA are considered to be useful for the diagnosis of several vasculitides, their significance for follow-up remains controversial and probably differs from one vasculitis to another.

We limit ourselves here to the importance of ANCA for diagnosis and outcome in the major systemic vasculitides, as the basic mechanisms of ANCA are discussed elsewhere.

24.2 Classification of Vasculitides

Vasculitides are a heterogeneous group of disorders, whose diverse and overlapping clinical manifestations as well as the type of blood vessels involved have prevented the development of a universally acceptable classification. Distinguishing subsets of vasculitides may be justified by the recognition of new symptoms that are not consistent with the usual presentation of the disease, differences in prognosis, requirements for different therapeutic strategies and unique pathogenic mechanisms. Finally, treatment may be optimized by application of specific therapeutic alternatives to each entity, based on the clinical symptoms, laboratory investigations and somewhat predictable outcome. This optimization should be the ultimate incentive for designation of subsets.

A detailed classification [7–10] and diagnostic criteria of the main types of vasculitides are given here. However, it is important at the outset to specify the differences between classical polyarteritis nodosa (PAN) and MPA. According to Fauci's classification [11], the PAN group comprises PAN, CSS and overlap angiitis. This pragmatic classification has been useful for clinicians but, recently, the recognition of a form of systemic vasculitides distinct from PAN, i.e. MPA, has been generally accepted. This distinction did not appear in the criteria for classification of PAN developed in 1990 by the American College of Rheumatology [7] (Table 24.1).

The reader should be aware that classification criteria are not appropriate to use for the diagnosis of individual patients but provide a standard way to evaluate and describe patients with vasculitis included in therapeutic trials or epidemiological

Table 24.1. 1990 ACR criteria for the classification of polyarteritis nodosa [7] (*BP* blood pressure, *BUN* blood urea nitrogen, *PMN* polymorphonuclear neutrophils)[a]

Criterion	Definition
1. Weight loss >4 kg	Loss of 4 kg or more of body weight since illness began, not due to dieting or other factors
2. Livedo reticularis	Mottled reticular pattern over the skin of portions of the extremities or torso
3. Testicular pain or tenderness	Pain or tenderness of the testicles, not due to infection, trauma, or other causes
4. Myalgias, weakness or polyneuropathy	Diffuse myalgias (excluding shoulder and hip girdle) or weakness of muscles or tenderness of leg muscles
5. Mononeuropathy or polyneuropathy	Development of mononeuropathy, multiple mononeuropathies, or polyneuropathy
6. Diastolic BP >90 mmHg	Development of hypertension with the diastolic BP higher than 90 mm Hg
7. Elevated BUN or creatinine	Elevation of BUN >40 mg/dl (14.3 mol/l) or creatinine >1.5 mg/dl (132 µmol/l), not due to dehydration or obstruction
8. Hepatitis B virus	Presence of hepatitis B surface antigen or antibody in serum
9. Arteriographic abnormality	Arteriogram showing aneurysms or occlusions of the visceral arteries, not due to arteriosclerosis, fibromuscular dysplasia, or noninflammatory causes
10. Biopsy of small- or medium-sized artery containing PMN	Histologic changes showing the presence of granulocytes or granulocytes and mononuclear leukocytes in the artery wall

[a] For classification purposes, a patient with vasculitis shall be said to have PAN if at least three of these ten criteria are present. The presence of any three or more criteria yields a sensitivity of 82.2% and a specificity of 86.6%.

studies. Recently, the Chapel Hill Consensus Conference Nomenclature [3] attempted to define a standardized classification system for some of the most common forms of noninfectious systemic vasculitides and to formulate basic definitions of these entities. The authors of the consensus took into consideration ten selected forms of vasculitis and assigned them to one of three categories (Table 24.2): large-sized vessel vasculitis, medium-sized vessel vasculitis or small-sized vessel vasculitis. The proposed distinguishing feature for PAN versus MPA (formerly microscopic polyarteritis) is the absence versus the presence of vasculitis in arterioles, venules or capillaries. Small-vessel involvement, when present, was the definitive diagnostic criterion of MPA and excluded the diagnosis of PAN even if medium-sized artery lesions were seen. This nomenclature has the advantage of emphasizing the existence of MPA but the disadvantage of giving it too large a place in the family of vasculitides. Histological findings of vasculitis, in nerve for instance, are often common to several necrotizing angiopathies. We now advocate taking into consideration, for PAN and MPA, classification parameters other than histopathology, for example manifestations especially of lung and kidney involvement, biological signs (ANCA, hepatitis B or hepatitis C virus infection) and angiographic data [12], and accepting the existence of an overlap syndrome.

24.2. Names and definitions adopted by the Chapel Hill Consensus Conference on the nomenclature of systemic vasculitis. *Large vessels* refers to the aorta and the largest branches directed toward major body regions (e.g. to the extremities and the head and neck); *medium-sized vessels* refers to the main visceral arteries (e.g. renal, hepatic, coronary, and mesenteric arteries); *small vessels* refers to venules, capillaries, arterioles, and the intraparenchymal distal arteries that connect with arterioles. Some small- and large-vessel vasculitides may involve medium-sized arteries, but large- and medium-sized vessel vasculitides do not involve vessels smaller than arteries. *Normal type* represents essential components; *italicized type* represents usual, but not essential, components

Large-vessel vasculitis	
Giant cell (temporal) arteritis	Granulomatous arteritis of the aorta and its major branches, with a predilection for the extracranial branches of the carotid artery. Often involves the temporal artery. Usually occurs in patients older than 50 years and often is associated with polymyalgia rheumatica
Takayasu's arteritis	Granulomatous arteritis of the aorta and its major branches. Usually occurs in patients younger than 50 years
Medium-sized vessel vasculitis	
Polyarteritis nodosa (classic polyarteritis nodosa)	Necrotizing inflammation of medium-sized or small arteries without glomerulonephritis or vasculitis in arterioles, capillaries, or venules
Kawasaki disease	Arteritis involving large, medium-sized, and small arteries, and associated with mucocutaneous lymph node syndrome. Coronary arteries are often involved. Aorta and veins may be involved. Usually occurs in children
Small-vessel vasculitis	
Wegener's granulomatosis[a]	Granulomatous inflammation involving the respiratory tract and necrotizing vasculitis affecting small- to medium-sized vessels (e.g. capillaries, venules, arterioles, and arteries). Necrotizing glomerulonephritis is common
Churg-Strauss syndrome[a]	Eosinophil-rich and granulomatous inflammation involving the respiratory tract, and necrotizing vasculitis affecting small- to medium-sized vessels, associated with asthma and eosinophilia
Microscopic polyangiitis (microscopic polyarteritis)[a]	Necrotizing vasculitis, with few or no immune deposits, affecting small vessels (i.e. capillaries, venules, or arterioles). Necrotizing arteritis involving small- and medium-sized vessels may be present. Necrotizing glomerulonephritis is very common. Pulmonary capillaritis often occurs
Henoch-Schönlein purpura	Vasculitis, with IgA-dominant immune deposits, affecting small vessels (i.e. capillaries, venules, or arterioles). Typically involves skin, gut, and glomeruli, and is associated with arthralgias or arthritis
Essential cryoglobulinemia vasculitis	Vasculitis, with cryoglobulin immune deposits, affecting small vessels (i.e. capillaries, venules, or arterioles), and associated with cryoglobulins in serum. Skin and glomeruli are often involved
Cutaneous leukocytoclastic angiitis	Isolated cutaneous leukocytoclastic angiitis without systemic vasculitis or glomerulonephritis

[a] Strongly associated with antineutrophil cytoplasmic autoantibodies

Table 24.3. Differential diagnosis of PAN and MPA

Criteria	PAN	MPA
Histology		
Type of vasculitis	Necrotizing with mixed cells, rarely granulomatous	Necrotizing with mixed cells, not granulomatous
Type of vessels involved	Medium- and small-sized muscle arteries, sometimes arterioles	Small vessels (i.e. capillaries, venules or arterioles). Small- and medium-sized arteries may be also affected
Distribution and localization		
Kidney		
Renal vasculitis with renovascular hypertension, renal infarcts and microaneurysms	Yes	No
Rapidly progressive glomerulonephritis	No	Very common
Lung		
Pulmonary hemorrhage	No	Yes
Peripheral neuropathy	50–80%	10–50%
Relapses	Rare	Frequent
Laboratory data		
ANCA	Rare (<20%)	Yes (50–80%)
HBV infection	Yes (uncommon)	No
Abnormal angiography (microaneurysms, stenoses)	Yes (variable)	No

24.2.1 Medium-Sized Vessel Vasculitides

PAN and Kawasaki disease are the two vasculitides affecting predominantly small-sized vessels mentioned in the Chapel Hill nomenclature [9]. We have detected ANCA in MPA and very rarely in PAN [13]. Perinuclear-labeling ANCA (p-ANCA) were observed in 74.5% of the cases and ELISA identified anti-myeloperoxidase (anti-MPO) antibodies in the majority of patients.

Conversely, cytoplasmic-labeling ANCA (c-ANCA) were rarely observed. At present, we consider that anti-MPO ANCA, when present, could be one of the criteria applied to discriminate between PAN and MPA. When ANCA are present, a diagnosis of PAN could not be considered and MPA could be diagnosed (Table 24.3)

24.2.2 Microscopic Polyangiitis

Although MPA was associated at the time of diagnosis with p-ANCA in 74.5% of a personal series of 85 patients [13] selected based on the Chapel Hill nomenclature criteria, their pathogenic role has not been fully elucidated in this disease. Jennette et al. [14] found the p-ANCA pattern in serum from about 50% of MPA patients with the antibodies often being directed against MPO. However, anti-MPO antibodies can also be detected in other systemic vasculitides as well as in some patients with rheumatoid arthritis, systemic lupus erythematosus or inflammatory bowel disease [15]. In addition, antibodies to proteinase 3 (PR3) may also be present in MPA, but less frequently, as was the case in four of our patients [13]. The detection of PR3 could evoke the diagnosis of WG but the long duration of follow-up for our four patients tended to reject this hypothesis and to confirm the diagnosis of MPA. The value of ANCA is high for MPA diagnosis and we think that ANCA-positivity should be considered diagnostic for this disease and exclusionary for PAN. Our observations suggest that ANCA may also be associated with a high relapse rate. However, this possibility remains to be confirmed in prospective trials.

24.2.3 Churg-Strauss Syndrome

CSS is also ANCA-associated and these antibodies are found in 50–66% of the patients [4, 15]. In a retrospective study [16], we have observed that many patients are ANCA-positive. p-ANCA are usually found and c-ANCA are rarely present. Very little information has been published on the follow-up of patients with ANCA-positive CSS [4] and no prospective study has demonstrated a relationship between ANCA persistence or reappearance and relapse. In a previous study [4], a few patients remained ANCA-positive during follow-up and relapsed several times. This outcome can also be observed in ANCA-negative CSS patients and it is too early to draw any conclusions on the necessity to intensify treatment or to reintroduce it when ANCA remain positive or reappear.

24.2.4 Wegener's Granulomatosis

More than 80% of patients [17] with systemic WG are ANCA-positive as are 50% of those with limited forms of the disease. The majority possess c-ANCA and ELISA detects anti-PR3 antibodies. Many studies have demonstrated the pathogenic role of anti-PR3 antibodies in the WG. p-ANCA, anti-MPO by ELISA, are rarely present. During follow-up, the persistence or reappearance of ANCA is more clearly associated with relapse. Indeed, Cohen-Tervaert [18] has observed that the ANCA titer increases before the recurrence of symptoms. Conversely, Kerr et al. [19] did not observe comparable events. Although a relationship between ANCA titer and relapse has been observed in some patients, others have not found such an evolution, and discrepancies between ANCA titer and clinical outcome have been noted. In a prospective study, we monitored ANCA during follow-up and observed that the persistence of ANCA was associated with relapse in every case (9/9) [20]. In 34/43 patients, ANCA disappeared. Of 13 patients who relapsed, 7 were ANCA-positive and 6 ANCA-negative. ANCA were also detected in 4/7 patients who had been ANCA-negative at diagnosis and remained so.

Therefore, at this time, it is difficult to draw any conclusions as to the relevance of ANCA during follow-up. When ANCA remain positive or when the titer increases, relapse seems probable, but the time to relapse varies widely and cannot be predicted. For patients in whom ANCA reappear, relapse may also occur, as it can in patients who remain ANCA-negative or in patients who were negative at the time of diagnosis.

ANCA play a major role in WG. They are of major importance for diagnosis and, in patients with clinical evidence of diagnosis, lung biopsy does not seem essential. In some cases, the reappearance of ANCA after their disappearance under treatment serves as a "warning" signal for an expected relapse at some time in the relatively near future.

24.3 Summary

ANCA are one of the hallmarks of several small-sized vessel vasculitides. Two major labeling patterns can be detected by immunofluorescence: perinuclear (p-ANCA) or diffuse cytoplasmic (c-ANCA), which, according to an ELISA correspond primarily to anti-MPO and anti-PR3 antibodies, respectively.

At disease onset, 80% of systemic WG and 50% of localized WG patients are ANCA-positive (mainly anti-PR3 c-ANCA). In MPA, 50–65% of patients are p-ANCA positive primarily anti-MPO ANCA.

ANCA contribute to the diagnosis of small-sized vessel vasculitis and can, in our opinion, discriminate between MPA and PAN. PAN, a predominantly medium-sized vessel vasculitis, is not ANCA-associated. ANCA should therefore be considered in diagnosing vasculitis.

At present, clinical studies have not established a close relationship between ANCA positivity and clinical symptoms or disease severity. Relapses are not associated with ANCA positivity at disease onset because relapses can also occur in ANCA-negative patients. In the case of a relapse, the irrelevance of ANCA has not yet been clearly established, except in WG, where the outcome might be poorer in ANCA-positive patients.

References

1. Van der Woude F, et al (1985) Autoantibodies against neutrophils and monocytes: tool for diagnosis and marker of disease activity in Wegener's granulomatosis. Lancet 1: 425–429
2. Kallenberg CG, et al (1992) Antineutrophil cytoplasmic antibodies: a still-growing class of autoantibodies in inflammatory disorders. Am J Med 93: 675–82
3. Cohen-Tervaert J, et al (1991) Antimyeloperoxidase antibodies in the Churg-Strauss syndrome. Thorax 46: 70–71
4. Cohen P, et al (1995) Persistence of antineutrophil cytoplasmic antibodies (ANCA) in asymptomatic patients with systemic polyarteritis nodosa or Churg-Strauss syndrome: follow-up of 53 patients. Clin Exp Rheumatol 13: 193–198
5. Guillevin L, et al (1990) Antineutrophil cytoplasmic antibodies in polyarteritis nodosa related to hepatitis B virus (letter). Arthritis Rheum 33: 1871–1872
6. Jennette JC, Falk RJ (1990) Antineutrophil cytoplasmic autoantibodies and associated diseases: a review. Am J Kidney Dis 15: 517–529
7. Lightfoot RJ, et al (1990) The American College of Rheumatology criteria for the classification of polyarteritis nodosa. Arthritis Rheum 33: 1088–1093
8. Masi A, et al (1990) The American College of Rheumatology criteria for the classification of Churg-Strauss syndrome (allergic granulomatosis angiitis). Arthritis Rheum 33: 1094–1100
9. Jennette JC, et al (1994) Nomenclature of systemic vasculitides. Proposal of an international consensus conference. Arthritis Rheum 37: 187–192
10. Leavitt R, et al (1990) The American College of Rheumatology criteria for the classification of Wegener's granulomatosis. Arthritis Rheum 33: 1101–1107
11. Fauci A, et al (1978) The spectrum of vasculitis: clinical, pathologic, immunologic and therapeutic considerations. Ann Intern Med 89: 660–676

12. Guillevin L, et al (1995) Antineutrophil cytoplasmic antibodies (ANCA) and abnormal angiograms in polyarteritis nodosa and Churg-Strauss syndrome: indications for the diagnosis of microscopic polyangiitis. Ann Méd Interne (Paris) 146: 548–550
13. Guillevin L, et al (1999) Microscopic polyangiitis: clinical and laboratory findings in 85 patients. Arthritis Rheum 42:421–430
14. Jennette JC, et al (1989) Antineutrophil cytoplasmic autoantibody-associated glomerulonephritis and vasculitis. Am J Pathol 135: 921–930
15. Hoffman G, Specks U (1998) Antineutrophil cytoplasmic antibodies. Arthritis Rheum 41: 1521–1537
16. Guillevin L, et al (1999) Churg-Strauss syndrome: clinical study and long-term follow-up of 96 patients. Medicine (Baltimore) 78:26–37
17. Gross WL (1995) Antineutrophil cytoplasmic autoantibody testing in vasculitides. Rheum Dis Clin North Am 21: 987–1011
18. Cohen-Tervaert J (1990) Prevention of relapses in Wegener's granulomatosis by treatment based on antineutrophil cytoplasmic antibody titer. Lancet 336: 709–711
19. Kerr GS, et al (1993) Limited prognostic value of changes in antineutrophil cytoplasmic antibody titers in patients with Wegener's granulomatosis. Adv Exp Med Biol 336: 411–414
20. Girard T, et al (1998). Antineutrophil cytoplasmic antibodies (ANCA) as a predictive factor for outcome in Wegener's granulomatosis (WG). A prospective study in 55 patients. Arthritis Rheum 41: S117

Chapter 25

Immunodiagnostic Aspects of Autoantibodies Against Myeloperoxidase

Elena Csernok, Wolfgang L. Gross

25.1 Introduction

The past two decades have seen an explosion of data on the new group of autoantibodies known collectively as antineutrophil cytoplasmic antibodies (ANCA) which are specific for antigens in the cytoplasm (mainly primary granules) of neutrophils and monocytes. These autoantibodies were originally detected by indirect immunofluorescence (IIF) on alcohol-fixed neutrophils and two different staining patterns were described: the cytoplasmic (classic) cANCA and the perinuclear pANCA (Fig. 25.1). cANCA is induced by autoantibodies directed against proteinase 3 (PR3) and pANCA is usually directed against myeloperoxidase (MPO) although other antigenic targets exist.

The so-called pANCA fluorescence pattern is an artifact of ethanol fixation caused by the translocation of positively charged proteins to the negatively charged nuclear membrane. If the fixation of cells is done with crosslinking fixatives (i.e. formalin), which prevent the perinuclear rearrangement of charged antigens, pANCA like cANCA exhibit a cytoplasmic staining. Because pANCA fluorescence can resemble that of antinuclear antibody (ANA), this has to be excluded by additional testing on a standard substrate for ANA, such as HEp2 cells.

The most important target antigen of pANCA is MPO, but autoantibodies against elastase, cathepsin G, lactoferrin, lysozyme, azurocidin, defensin, human lysosome-associated membrane protein 2, β-glucuronidase, catalase, α-enolase and bactericidal permeability-increasing protein have all been identified as causing the pANCA fluorescence pattern. In a significant proportion of pANCA-positive sera, the target antigens remain to be identified. Thus, the list of target antigens of ANCA will continue to grow.

ANCA immunofluorescence patterns and antigen specificities are associated with a wide disease spectrum. The discovery that ANCA are strongly associated with vasculitic disorders has changed dramatically the diagnostic approach to patients with clinically suspected vasculitis and/or glomerulonephritis. Antibodies against PR3 (PR3-ANCA) are mainly detected in serum from patients with Wegener's granulomatosis (WG). Likewise, patients with microscopic polyangiitis (MPA) and necrotizing crescentic glomerulonephritis (NCGN) have a high incidence of pANCA antibodies directed against MPO (MPO-ANCA) [1].

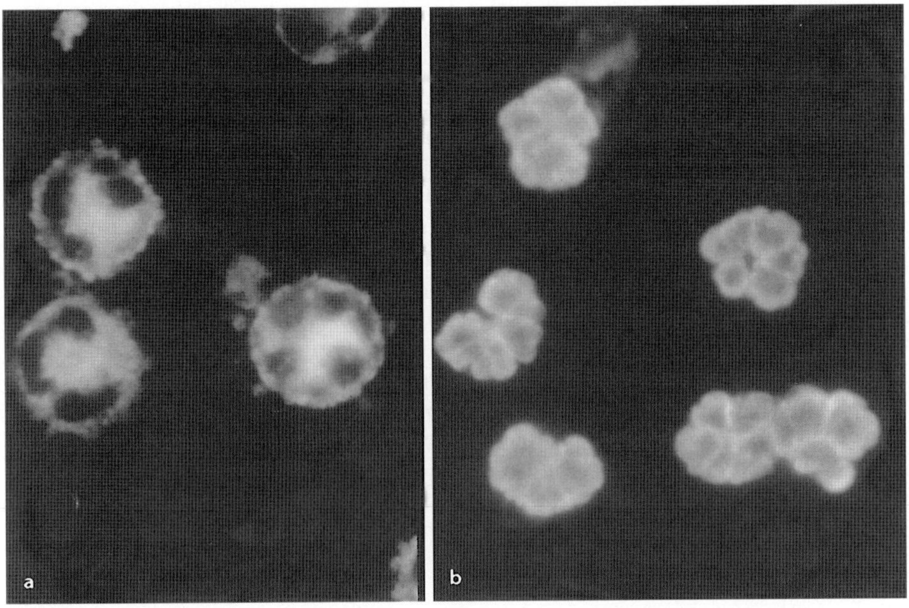

Fig. 25.1 a,b. ANCA patterns shown by IIF. **a** Fine granular cANCA pattern; **b** pANCA pattern

25.2 Methods of ANCA Detection

Several methods are used to detect ANCA: IIF, ELISA, radioimmunoassay, immunoblotting, immunoprecipitation. The first method used to determine ANCA was IIF and it remains the "gold standard" of ANCA screening. The reliability of IIF assays depends on the type of substrate employed, the source of the cells (human neutrophils, cell lines such as HL-60), fixation, storage, incubation and washing steps. To improve the specificity and to avoid problems associated with IIF (i.e. lack of antigen specificity, problems with the interpretation of IIF patterns) solid-phase immunoassays with highly purified target antigens (ELISA) have been applied. For pANCA differentiation antigen-specific ELISAs are used in many laboratories. Depending on the degree of purification of the antigen used (native purified or recombinant proteins), ELISA may be used as a complementary method to IIF in order to determine the specificity of ANCA, e.g. PR3-ANCA, MPO-ANCA, BPI-ANCA, etc.

25.3 Sensitivity and Specificity of Various Commercial MPO-ANCA ELISA Methods

While extensive European multicenter studies have characterized and standardized the performance of PR3- and MPO-ANCA [2], very little data about antigen purity, analytic sensitivity and specificity are available about ELISA kits offered commercially in Europe and in the United States.

In recent years a growing number of reports have dealt with diseases related to the presence of PR3-/MPO-ANCA in serum. One puzzling fact that emerges from the literature is the variability of the prevalence of these autoantibodies in diseases known to be associated with ANCA. Although many commercial kits are now available for the detection of PR3-/MPO-ANCA, there is no uniform international methodology. Recently, we compared eight commercial ELISA kits and an in-house method (HM) in terms of their specificity and sensitivity in detecting MPO-ANCA. Sera from healthy controls, four out of five SLE and all pANCA-negative sera were found to be MPO-ANCA-negative with all eight ELISA kits. In the group of pANCA-positive sera, the MPO-ANCA positivity ranged from 25% to 75%. However, only 35% of MPO-ANCA-positive sera were confirmed by capture ELISA, immunoblotting and inhibition assay. The concordance rate was only 30% among MPO-ANCA-positive sera with the eight ELISA kits. No significant correlation was observed between pANCA and MPO-ANCA levels. The HM ELISA showed that 45% of pANCA sera were also positive for MPO-ANCA. Our results indicate that the sensitivity and specificity for MPO-ANCA detection differ significantly among the ELISA kits tested and underline the necessity of establishing uniform international standards for ANCA ELISA procedures, in order to permit more reliable interpretation and comparison of data [3].

25.4 Disease Associations of MPO-ANCA

In contrast to PR3-ANCA, MPO-ANCA are not specific for one disease entity within the spectrum of vasculitides, but occur in a variety of disorders which share several common clinical and histological features. MPO-ANCA are found in: (1) systemic vasculitides such as MPA (40–60%), WG (3%), Churg-Strauss syndrome (CSS) (5%) and rarely in secondary vasculitis; (2) NCGN (approximately equivalent to renal vasculitis) (the specificity is rather high: 40–70%) and rarely in other forms of non-vasculitic glomerulonephritis; (3) chronic inflammatory rheumatic disorders, e.g. rheumatoid arthritis (3%); (4) collagen vascular diseases such as SLE (5–40%) [4–8]. Additionally, MPO- ANCA can be induced by drugs (hydralazine, propylthiouracil, l-tryptophan) which can induce diseases associated with vasculitis (e.g. glomerulonephritis) and pulmonary vasculitis [9]. Occupational exposure to environmental factors such as silica dust may provoke MPO-ANCA-associated rapidly progressing glomerulonephritis (RPGN) [10].

In 30% of patients suffering from Goodpasture's syndrome, MPO-ANCA were detected in addition to antibodies directed against the glomerular basement membrane [11]. Furthermore, MPO-ANCA have been reported in infective disorders [12].

The significance of following titers of MPO-ANCA in patients with MPO-ANCA-associated diseases has not been studied in detail. Preliminary data suggest that titers of the autoantibodies parallel disease activity [13].

25.5 Sensitivity and Specificity of MPO-ANCA Test for the Diagnosis of Vasculitis

A recently published report has defined the disease associations and the sensitivity/specificity of PR3- and MPO-ANCA obtained in a multicenter study supported by the European Union. The aim of this study was to evaluate the diagnostic value of ANCA detection by IIF and antigen-specific ELISA performed in different locations, in patients with idiopathic small-vessel vasculitides. A total of 14 centers participated in the standardization study on ANCA assays and a total of 300 patients with idiopathic systemic vasculitides or RPGN were investigated. The results were compared with 184 disease controls and 740 healthy controls. The sensitivity and specificity were follows: in MPA sensitivity for pANCA was 58% and for MPO-ANCA 58%; in RPGN for pANCA 45% and for MPO-ANCA 64%. The specificity of assays (related to disease controls) were for pANCA 81% and for MPO-ANCA 91%. When the results of IIF were combined with those of ELISA the diagnostic specificity increased to 99% [14].

25.6 Clinical Spectrum Associated with MPO-ANCA Positivity

Studies that have compared the clinical and histopathological associations between patients' PR3-ANCA and MPO-ANCA are very few. In most of them, patient selection criteria have included the presence of a renal biopsy showing NCGN. In a recent study, Franssen et al. compared for the first time the spectrum of PR3-ANCA- and MPO-ANCA-associated diseases by analyzing the clinical/histological features and outcome of consecutive ANCA-positive patients who were not selected for the presence of any particular organ manifestation [15]. They found that at the time of diagnosis, patients with PR3-ANCA had a higher vasculitis activity index and that the mean number of affected organs in the PR3-ANCA group exceeded that in the MPO-ANCA group. In particular, extrarenal manifestation, granuloma and relapses were more prominent in PR3-ANCA- than in MPO-ANCA-positive patients [15]. This large divergence in the disease spectrum between patients with PR3-ANCA and those with MPO-ANCA needs to be confirmed by other groups.

25.7 Conclusion

MPO-ANCA are not specific for one disease entity within the spectrum of vasculitides, but occur in a variety of disorders which share several common clinical and histological features. These autoantibodies are particularly important serological markers of certain forms of small vessel vasculitides, such as MPA and NCGN. Before including MPO-ANCA test results in the clinical decision-making process, there is the necessity of establishing uniform international standards for ANCA-ELISA procedures in order to permit more reliable interpretation and comparison of data. MPO-ANCA test results must be carefully interpreted in conjunction with clinical, laboratory and histological findings.

25.8 Summary

The diagnosis and classification of primary systemic vasculitides were considerably advanced by the discovery and characterization of ANCA, a new family of autoantibodies directed against various components of the neutrophil cytoplasm. The close association of certain ANCA subspecificities, such as PR3- and MPO-ANCA, with WG and MPA, respectively, has led to their designation as "ANCA-associated vasculitis".

In contrast to PR3-ANCA, MPO-ANCA are not specific for one disease entity within the spectrum of vasculitides, but occur in a variety of disorders which share several common clinical and histological features. Their specificity for the group of idiopathic vasculitides including idiopathic NCGN is rather high (40–70%). MPO-ANCA occur in the majority of patients with MPA and idiopathic NCGN. Furthermore, they are found in a minority of CSS and WG patients (who are negative for PR3-ANCA). Autoantibodies against MPO are also detected in Goodpasture's syndrome (30–40%) and in hydralazine-induced glomerulonephritis, both conditions known to be associated with small-vessel vasculitis. The reported prevalence of MPO-ANCA in systemic lupus erythematosus (SLE) and rheumatoid arthritis (RA) differs between several studies.

The growing clinical interest in the determination of ANCA has led many manufacturers to develop antigen-specific ELISAs for assessing these autoantibodies in serum and other biological fluids. Recently, we compared different commercial kits in terms of their specificity and sensitivity. Our results demonstrate that the specificity and sensitivity of MPO-ANCA detection differ markedly among the ELISA kits tested. Furthermore, the different sensitivities and low concordances of ANCA detection with the ELISA kits could partly explain the differences reported in the literature. Moreover, the results underline the necessity of establishing uniform international standards for ANCA ELISA procedures in order to permit more reliable interpretation and comparison of data. The sensitivity and specificity of MPO-ANCA tests for the diagnosis of vasculitides were evaluated in a multicenter study supported by the European Union and were follows: in MPA sensitivity for pANCA was 58% and for MPO-ANCA 58%; in idiopathic NCGN for pANCA 45% and for MPO-ANCA 64%. The specificity of assays (related to disease controls) were for pANCA 81% and for MPO-ANCA 91%. When the results of IIF were combined with those of ELISA the diagnostic specificity increased to 99%.

In conclusion, MPO-ANCA are particularly important serological markers for certain forms of small-vessel vasculitides, such as MPA and NCGN. MPO-ANCA test results must be carefully interpreted in conjunction with clinical, laboratory and histological findings.

References

1. Gross WL, Csernok E (1995) Immunodiagnostic and pathophysiologic aspects of antineutrophil cytoplasmic antibodies in vasculitis. Curr Opin Rheumatol 7: 11–19
2. Hagen EC, Andrassy K, Csernok E, Daha MR, Gaskin G, Gross WL, Hansen B, Heigl Z, Hermans J, Jayne D, Kallenberg CGM, Lesavre Ph, Lockwood CM, Lüdemann J, Mascart-Lemone F, Mirapeix E, Pusey CD, Rasmussen N, Sinico RA, Tzioufas A, Wieslander J, Wiik A, Van der Woude FJ (1996) Development and standardization of solid phase assays for the detection of anti-neutrophil cytoplasmic antibodies (ANCA). A report on the second phase of an international study on the standardization of ANCA assays. J Immunol Methods l96: 1–15
3. Wang G, Csernok E, Groot K de, Gross W (1997) Comparison of eight commercial kits for quantitation of antineutrophil cytoplasmic antibodies (ANCA). Immunol Methods 208: 203–211
4. Gueirard P, Delpech A, Gilbert D, et al (1991) Anti-myeloperoxidase antibodies: immunological characteristics and clinical associations. J Autoimmun 4: 517–527
5. Ulmer M, Rautmann A, Gross WL (1992) Immunodiagnostic aspects of autoantibodies against myeloperoxidase. Clin Nephrol 37: 161–168
6. Hauschild S, Schmitt WH, Csernok E, Flesch BK, Rautmann A, Gross WL (1992) ANCA in systemic vasculitides, collagen vascular diseases, rheumatic disorders and inflammatory bowel diseases. Adv Exp Med Biol 336: 245–251
7. Braun MG, Csernok E, Schmitt WH, Gross WL (1996) Incidence, target antigens, and clinical implications of antineutrophil cytoplasmic antibodies in rheumatoid arthritis. J Rheumatol 23: 826–830
8. Schnabel A, Csernok E, Isenberg DA, Mrowka C, Gross WL (1996) Antineutrophil cytoplasmic antibodies in systemic lupus erythematosus. Arthritis Rheum 38: 633–637
9. Nässberger L, Sjöholm AG, Jonsson H, et al (1993) Autoantibodies against neutrophil cytoplasm components in systemic lupus erythematosus and in hydralazine-induced lupus. Clin Exp Immunol 81: 380–383
10. Niles JL, Pan G, Collins AB, et al (1991) Antigen-specific radio- immunoassay for anti-neutrophil cytoplasmic antibodies in the diagnosis of rapidly progressive glomerulonephritis. J Am Soc Nephrol 2: 27–36
11. Jayne DRW, Marshall PD, Jones SJ, Lockwood CM (1990) Autoantibodies to GBM and neutrophil cytoplasm in rapidly progressive glomerulonephritis. Kidney Int 37: 965–970
12. Gallichio MC, Savige JA (1991) Detection of antimyeloperoxidase and anti-elastase antibodies in vasculitis and infections. Clin Exp Immunol 84: 232–237
13. Cohen-Tervaert JW, Goldschmeding R, Elema JD, et al (1990) Association of autoantibodies to myeloperoxidase with different forms of vasculitis. Arthritis Rheum 33: 1264–1272
14. Hagen EC, Daha MR, Hermans J, Andrassy K, Csernok E, Gaskin G, Lesavre P, Lüdemann J, Rasmussen N, Sinico RA, Wiik A, Woude F van der (1998) Diagnostic value of standardized assays for antineutrophil cytoplasmic antibodies in idiopathic systemic vasculitis. Kidney Int 53: 743–753
15. Franssen C, Gans R, Kallenberg C, Hageluken C, Hoorntje S (1998) Disease spectrum of patients with antineutrophil cytoplasmic autoantibodies of defined specificity. Distinct differences between patients with anti-proteinase 3 and anti-myeloperoxidase autoantibodies. J Intern Med 244: 209–216

Chapter 26

Pathophysiological Mechanisms in Anti-Myeloperoxidase-Associated Vasculitis

Jan Willem Cohen Tervaert

26.1 Introduction

Antibodies to myeloperoxidase (MPO) were first reported during the first ANCA workshop in January 1988 in patients with either Wegener's granulomatosis [1, 2] or idiopathic necrotizing crescentic glomerulonephritis [2]. Anti-MPO antibodies were later shown to be present also in patients with microscopic polyangiitis or Churg-Strauss syndrome (Table 26.1) [3–6]. Besides being a helpful diagnostic tool, determination of anti-MPO levels can also be useful for monitoring disease activity, since relapses of disease are often preceded by rises in anti-MPO levels [5, 7]. Furthermore, persisting high levels of anti-MPO are associated with a poor outcome of renal disease [8]. The strong relationship between anti-MPO and primary vasculitic syndromes suggests an important role for anti-MPO in the pathophysiology. Experimental evidence for the pathogenic potential of anti-MPO has been derived from both in vitro and in vivo studies and is reviewed.

26.2 Pathophysiology of Anti-MPO-Associated Vasculitis: In Vitro Data

In vitro, anti-MPO antibodies can activate neutrophils primed with tumor necrosis alpha (TNF-α) for the production of reactive oxygen intermediates (ROI), the release of lysosomal enzymes, and the secretion of interleukin-1ß [9, 10]. Furthermore, it has been demonstrated that anti-MPO antibodies are able to stimulate neutrophils to adhere to cultured human endothelial cells, a process that can be inhibited by anti-CD18 antibodies [11]. More recently, Johnson et al. have clarified the mechanism by which anti-MPO might stimulate neutrophil adherence by showing that anti-MPO IgG fractions stimulate the upregulation of CD11b on neutrophils in vitro [12]. Finally,

Table 26.1. Characteristics of ANCA-associated vasculitides (*NCGN* necrotizing crescentic glomerulonephritis; ++++ >70% of patients, +++ 30–70% of patients, ++ 10–30% of patients, + <10% of patients)

Disease	Clinical	Anti-MPO antibody	Anti-PR3 antibody
Churg-Strauss syndrome	Asthma, eosinophilia, neuropathy	++++	+
Wegener's granulomatosis	Nose bleeds, nephritis, lung lesions	++	++++
Microscopic polyangiitis	Nephritis, purpura, hemoptysis	+++	+++
Idiopathic NCGN	nephritis	++++	++

it has been demonstrated that anti-MPO-stimulated primed neutrophils can lyse cytokine-pretreated cultured endothelial cells [13, 14]. It is still a matter of debate whether anti-MPO antibodies can also activate monocytes [15] and/or endothelial cells [16]. One should realize that trace amounts of endotoxin could be present in the isolated IgG fractions and hence may cause the observed activation of these cells.

The mechanisms involved in anti-MPO-mediated neutrophil activation are not completely understood. Upon priming with TNF-α, neutrophils express MPO on the cell surface which then become accessible for interaction with anti-MPO [9]. It is thought that MPO binding to the cell membrane is predominantly through charge interactions, since no receptor for MPO has been identified. Binding of ANCA to primed neutrophils may result in activation of neutrophils, a process that is largely dependent on engagement of $\beta 2$ integrins and the interaction of the Fc portion of anti-MPO [17]. A Fc-independent mechanism, however, has also been described as being operative in vitro [9, 18].

26.3 Pathophysiology of Anti-MPO-Associated Vasculitis: Animal Studies

Several types of pathophysiological event that may lead to vessel wall damage in vasculitis are currently recognized. These include pathogenic immune complex deposition or in situ formation, a "Shwartzman-like" phenomenon in which intravascular activation and aggregation of neutrophils may be operative, antibody-dependent cell-mediated cytotoxicity, and cell-mediated immune responses [19]. Pathogenic immune complex (ICX) deposition-mediated vasculitis is best depicted in the serum sickness animal model. In this model, rabbits are injected with bovine serum albumin and about 7–10 days later immune complexes are found that may induce vasculitis and/or glomerulonephritis. In anti-MPO-associated vasculitis, however, immune complexes are generally not found in the lesions. Therefore, the classic renal lesion in anti-MPO glomerulonephritis is labelled "pauci-immune". In kidney biopsies of patients with anti-MPO-mediated glomerulonephritis, we have found no IgA or IgG deposits and only nonspecific IgM deposits in a minority of patients [20]. Complement deposition, however, is often present (in about 50% of the cases) [20]. This may point to prior ICX deposition, but there is no proof for this hypothesis.

To test the hypothesis that anti-MPO itself may induce vasculitis, we immunized BN rats with human MPO which induced antibodies to human MPO which also cross-reacted with rat MPO [21]. Furthermore, in these rats a cellular response to MPO could be detected [21]. To our surprise these rats appeared completely normal and no vasculitic lesions were found at autopsy. So, the induction of anti-MPO antibodies is not sufficient to induce vasculitis in rats. We hypothesized that there must first be ICX deposition at vessel walls. These ICX then attract neutrophils and these neutrophils then express MPO on their cell surface [19] that may bind anti-MPO which results in an overstimulation of the neutrophils resulting in vasculitis and also the rapid disappearance of ICX. To test this hypothesis, we injected MPO-immunized rats with an extract of neutrophils, containing MPO, and hydrogen peroxide (H_2O_2). In this context we predicted that ICX deposition and subsequently vasculitis would occur. We observed vasculitis of the lungs and the gut in the rats that were immunized with MPO but not in the rats that were not MPO immunized and received the neutrophil extract only [22]. Unfortunately, no glomerulonephritis was found. However, after unilateral perfusion of the left kidney with the neutrophil extract and H_2O_2 we saw a severe form of necrotizing crescentic glomerulonephritis in the rats that had been immunized with MPO and no lesions in non-immunized rats [21]. More importantly, immediately after perfusion, ICX deposits were seen in the kidneys, but these were gone within a very short period and when the glomerulonephritis was at his maximum no further immune deposits were detected [21].

So, in the presence of anti-MPO severe vasculitis and pauci-immune glomerulonephritis can be induced in rats when immune complexes are first deposited along the vessel wall. The next question was what would happen if ICX other than anti-MPO/MPO ICX are deposited along the vessel wall. To study this, Heeringa et al. injected rats with an antibody to the rat glomerular basement membrane (GBM) and compared MPO-immunized rats with non-immunized rats. For these studies, a low dose of anti-GBM antibody was used that binds to the GBM but itself is not enough to induce a glomerulonephritis [23]. In rats with anti-MPO ANCA a severe glomerulonephritis developed whereas no lesions were found in the non-immunized rats.

From these experiments, we come to our current working hypothesis. Initially, there is ICX deposition and any antigen can be involved. Normally, this ICX deposition does not result in disease but in the presence of ANCA a severe glomerulonephritis or vasculitis may develop.

26.4 Induction of Autoimmunity to MPO

The central mechanism in autoimmunity is the breaking of self-tolerance. It is now well established that autoreactive T and B cells exists in the blood of healthy individuals and that these cells can potentially induce autoimmunity if activated beyond a certain threshold [24]. A combination of risk factors may be present in patients who develop autoimmunity to MPO. Inherited determinants have been sought, i.e. associations with certain HLA class I or class II molecules, Fc-gamma RIIa polymorphism and/or with the C3F component of complement, but have not been convincingly found in patients with anti-MPO vasculitis and/or glomerulonephritis. Environmental factors, however, are probably important modulators. Among these, drugs such as propylthiouracil and/or hydralazine [25] and/or exposure to chemical substances such as silicon [26] have been incriminated. Infectious agents are, however, the most likely candidates causing autoimmunity. Several mechanisms by which infectious agents might induce autoimmunity have been postulated [27]. These include molecular mimicry, abnormal presentation of self proteins, and/or abnormal stimulation of autoreactive T or B cells by agents such as superantigens. This latter mechanism has our special attention, since superantigens produced by *Staphylococci* that may activate autoreactive B cells, in a T-cell dependent way, to produce anti-MPO are often present in patients with anti-MPO-mediated vasculitis.

26.5 Summary

Vascular damage in anti-MPO-mediated vasculitis and/or glomerulonephritis results predominantly from activation of neutrophils when these cells adhere to endothelial cells. In situ-formed ICX may initially attract these neutrophils. Once these adherent activated neutrophils are "over-stimulated" by anti-MPO they may cause necrotizing vasculitis and glomerulonephritis and, in addition, stimulate the rapid disappearance of ICX, thus explaining the absence of ICX in tissue biopsies.

References

1. Cohen Tervaert JW, Huitema MG, Giessen M van der, Goldschmeding R, Woude FJ van der, Kallenberg CGM (1989) Wegener's granulomatosis and anticytoplasmic antibodies: the Groningen experience. APMIS 97: (S6) 36
2. Falk RJ, Jennette JC (1989) Immunofluorescence and ELISA determination of ANCA with description of a sub-class with anti-myeloperoxidase activity. APMIS 97: (S6) 45
3. Falk RJ, Jennette JC (1988) Anti-neutrophil cytoplasmic autoantibodies with specificity for myeloperoxidase in patients with systemic vasculitis and idiopathic necrotizing and crescentic glomerulonephritis. N Engl J Med 318: 1651–1657
4. Cohen Tervaert JW, Goldschmeding R, Elema JD, Giessen M van der, Huitema MG, Hem GK van der, Borne AEGKr von dem, Kallenberg CGM (1990) Autoantibodies against myeloid lysosomal enzymes in crescentic glomerulonephritis. Kidney Int 37: 799–806
5. Cohen Tervaert JW, Goldschmeding R, Elema JD, Giessen M van der, Huitema MG, Koolen MI, Hene RJ, The TH, Hem GK van der, Borne AEGKr von dem, Kallenberg CGM (1990) Association of autoantibodies to myeloperoxidase with different forms of vasculitis. Arthritis Rheum 33: 1264–1272
6. Cohen Tervaert JW, Limburg PC, Elema JD, Huitema MG, Horst G, The TH, Kallenberg CGM (1991) Autoantibodies against myeloid lysosomal enzymes: a useful adjunct to classification of patients with biopsy-proven arteritis. Am J Med 91: 59–66
7. Cohen Tervaert JW (1996) The value of serial ANCA testing during follow-up studies in patients with ANCA-associated vasculitides. A review. J Nephrol 9: 232–240
8. Franssen CFM, Stegeman CA, Oost-Kort WW, Kallenberg CGM, Limburg PC, Tiebosch A, Jong PE de, Cohen Tervaert JW (1998) Determinants of renal outcome in anti-myeloperoxidase associated necrotizing crescentic glomerulonephritis. J Am Soc Nephrol 9: 1915–1923
9. Falk RJ, Terell RS, Charles LA, Jennette JC (1990) Anti-neutrophil cytoplasmic autoantibodies induce neutrophils to degranulate and produce oxygen radicals in vitro. Proc Natl Acad Sci U S A 87: 4115–4119
10. Brooks CJ, King WJ, Radford DJ, Adu D, McGrath M, Savage COS (1996) IL-1β production by human polymorphonuclear leukocytes stimulated by anti-neutrophil cytoplasmic autoantibodies: relevance to systemic vasculitis. Clin Exp Immunol 106: 273–279
11. Ewert BH, Becker ME, Jennette JC, Falk RJ (1995) Antimyeloperoxidase antibodies induce neutrophil adherence to cultured human endothelial cells. Renal Failure 17: 125–133
12. Johnson PA, Alexander HD, McMillan SA, Maxwell AP (1997) Up-regulation of the granulocyte adhesion molecule Mac-1 by autoantibodies in autoimmune vasculitis. Clin Exp Immunol 107: 513–519
13. Ewert BH, Jennette JC, Falk RJ (1992) Anti-myeloperoxidase antibodies stimulate neutrophils to damage human endothelial cells. Kidney Int 41: 375–383
14. Savage COS, Pottinger BE, Gaskin G, Pusey CD, Pearson JD (1992) Autoantibodies developing to myeloperoxidase and proteinase 3 in systemic vasculitis stimulate neutrophil cytotoxicity towards cultured endothelial cells. Am J Pathol 141: 335–342
15. Casselman BL, Kilgore KS, Miller BF, Warren JS (1995) Antibodies to neutrophil cytoplasmic antigens induce monocyte chemoattractant protein-1 secretion from human monocytes. J Lab Clin Med 126: 495–502
16. Johnson PA, Alexander HD, McMillan SA, Maxwell AP (1997) Up-regulation of the endothelial cell adhesion molecule intercellular adhesion molecule-1 (ICAM-1) by autoantibodies in autoimmune vasculitis. Clin Exp Immunol 108: 234–242

17. Reumaux D, Vossebeld PJM, Roos D, Verhoeven AJ (1995) Effect of tumor necrosis factor-induced integrin activation on Fc-gamma receptor II-mediated signal transduction: relevance for activation of neutrophils by anti-proteinase 3 or anti-myeloperoxidase antibodies. Blood 86: 3189–3195
18. Kettritz R, Jennette JC, Falk RJ (1997) Crosslinking of ANCA-antigens stimulates superoxide release by human neutrophils. J Am Soc Nephrol 8: 386–394
19. Cohen Tervaert JW, Kallenberg CGM (1996) The role of autoimmunity to myeloid lysosomal enzymes in the pathogenesis of vasculitis. In: Hansson GK, Libby P (eds) Immune functions of the vessel wall. Harwood Academic Publishers, London, pp 99–120
20. Brouwer E, Huitema MG, Mulder AHL, Heeringa P, Goor H van, Cohen Tervaert JW, Weening JJ, Kallenberg CGM (1994) Neutrophil activation in vitro and in vivo in Wegener's granulomatosis. Kidney Int 45: 1120–1131
21. Brouwer E, Huitema MG, Klok PA, Cohen Tervaert JW, Weening JJ, Kallenberg CGM (1993) Antimyeloperoxidase-associated proliferative glomerulonephritis: an animal model. J Exp Med 177: 905–914
22. Heeringa P, Foucher P, Klok PA, Huitema MG, Cohen Tervaert JW, Weening JJ, Kallenberg CGM (1997) Systemic injection of products of activated neutrophils and H_2O_2 in myeloperoxidase-immunized rats leads to necrotizing vasculitis in the lungs and gut. Am J Pathol 151: 131–140
23. Heeringa P, Brouwer E, Klok PA, Huitema MG, Born J van den, Weening JJ, Kallenberg CGM (1996) Autoantibodies to myeloperoxidase aggravate mild anti-glomerular-basement-membrane mediated glomerular injury. Am J Pathol 149: 1695–1706
24. Kotb M (1995) Short analytical review. Infection and autoimmunity: a story of the host, the pathogen, and the copathogen. Clin Immunol Immunopathol 74: 10
25. Merkel PA (1998) Drugs associated with vasculitis. Curr Opin Rheumatol 10: 45
26. Cohen Tervaert JW, Stegeman CA, Kallenberg CGM (1998) Silicon exposure and vasculitis. Curr Opin Rheumatol 10: 12–17
27. Cohen Tervaert JW, Popa ER, Bos NA (1999) The role of superantigens in vasculitis. Curr Opin Rheumatol 11: 24–33

Subject Index

ACE 80
Acroleine 104
Adaptin 12
ADCC 162
ADVIA 136
Albumin 103
Aldehyde, reactive 104
ALL 151
Alzheimer disease 98
Amiloride 109
AML 151, 154
ANCA 173, 180
APL-M3 95
ApoE 98
ARF 11
ARNO3 11
Asthma 75, 98
Atherosclerosis 73, 81
Autophagocytosis 17
Azide 62
Azurocidin 23

Bacteria 53
Bayer 136, 142, 157
Bigenic model 141
Bromide 4, 59, 75, 129

calcium 32
Calneurin 25, 45
Calreticulin 25, 45
Candida 7, 41, 68, 152
Cardiovascular disease 137
Catalase 3, 62, 75, 77, 102
Cathepsin 23, 28, 29
Cell line 32 D 23
Cell line 985 46
Cell line CHO 38
Cell line HEP2 180
Cell line Hl60 16, 18, 27, 28
Cell line K562 126
Cell line K562 48
Cell line NB4 29
Cell line RBL1 23, 29
Chaperone 25, 28, 45
Chloramine 6, 52, 109, 110
Chloride 4, 5, 33, 41, 52, 58, 102, 159
3-chlorotyrosine 6, 81, 104, 105
Chronic granulomatous disease 54, 162
Churg Strauss syndrome 173, 177
Clathrin 12, 15
Complement receptor type I 29, 111
Compound heterozygosity 165
Compound I 5, 34, 58, 59, 65, 102, 130
Compound II 59, 65
Compound III 60, 61
COP I 12
COP II 11
CRT 50
Cystic fibrosis 81, 107

Defensins 7, 22, 28, 109
Diabetes mellitus 98, 150, 151
Diiodotyrosine 130
Dityrosine 103

E.coli 41.52
EDRF 72
Elastase 23, 109
Endocytosis 14
Endoglycosidase H 25
Endoperoxide synthase 31
Endosome 15
Eosinophil peroxidase 75, 114
–, deficiency 122, 125, 136, 142
–, dogs
–, recombinant 126
Epsin 13
ERAD 49
ERP 72 45

Fluoride 4
FMLP 160
fucose 32

Gaucher disease 162
Gelatinase 22, 23, 109
Glucose oxidase 55
ß-glucuronidase 29
glutathione reductase 81
GM-CSF 42, 70, 159
Golgi-apparatus 16, 23, 27
Granule 18, 22, 23
Granzyme 23

Heme 32, 33, 34, 35
Hemoglobin 117
Histidine 31
HIV 41
Horseradish peroxidase 3, 75
HSP 90 45
Hydrogen peroxide 5, 66, 102, 109
4-hydroxyphenylacetic acid 75
hypochlorite 52, 53
Hypochlorous acid 5, 52, 58, 59, 66, 102, 109
Hypothiocyanite 59
Hypothyroidism 133

I-cell disease 16
Inflammatory bowel disease 73
Interleukin 8 109
Iodide 4, 52, 59, 129
Iron deficiency 150, 151
Iron sulfur center 53

Kawasaki 177
KEDL 50
KFERQ-protein 17
Klebanoff 4

Lactoferrin 29, 117
Lactoperoxidase 75
–, gene polymorphisms 118
–, heme active site 116
–, heterogeneity 117
–, substrate binding site 118
–, three dimensional structure 116

LDL oxidation 105
Lead intoxication 150, 151
Leukemia 86, 93
LUC 136
Lung injury 73
Lysosome 11
Lysozyme 7, 22

Macrophage 42, 68
Manganese superoxide dismutase 81
mannose 32
Mannose-6-phosphate receptor 14, 15
MDS 151
methionine 40
Microglia 97, 99
Monensin 16
Monocytes, MPO deficiency 141
MPA 173
MPXI 137, 158, 160
multiple sclerosis 81, 93, 96
Mycobacterium 41
Myeloperoxidase, deletion mutant 26, 38
–, acute leukemia 86, 93
–, ANCA 180, 181
–, atherosclerosis 105
–, autoantibodies 42, 188
–, biochemistry 5, 52
–, biosynthesis 45
–, bronchial secretions 110
–, cystic fibrsois 107
–, deficiency 7, 48, 54, 122, 125, 136, 145, 147, 148, 150, 152, 161, 162, 166, 171
–, genetics 141, 168
–, heme binding site 40, 41, 116
–, historical aspects 1
–, macrophage 97
–, mannose content 39
–, microglia 97
–, monoclonal antibody 86, 154, 176, 186
–, processing 25, 45
–, purification 2
–, reaction mechanism 58
–, recombinant 38, 68
–, redox transformation 58
–, staining 168
–, superfamily 114
–, three diemnsional structure 116
–, tumor surveillance 153
–, x-ray cristallography 31
Myeloproliferative diseases 151
Myocardial infarction 73

Na/H-exchanger 109
NADPH oxidase 5
Neuroblastoma 162
Neurodenegeration 73
Neutrophil gelatinase associated lipocalin (NGAL) 28
Nitrate 73, 74
Nitric oxide 72
Nitric oxide synthase 72
Nitrite 52, 58, 61, 64, 73, 75
Nitrogen dioxide 61
Nitrotyrosine synthase 78, 79
3-nitrotyrosine 62, 73, 76, 80
Nitrylchloride 74
NSF 13

o-dianisidine 39
OriC 55
Osteomyelitis 162
Oxyhemoglobin 73
OxyMPO 60
Oxymyoglobin 73

P1 phage 55
Perchlorate discharge test 133
Perox-diagram 136, 158
Peroxidase cycle 78
Peroxynitrite 73
Phagosome 63
Phenylketonuria 162
Phorbol myristate acetate 76
p-hydroxyphenylacetalaldehyde 104
Plasmid 55
Porphyria 162
Presentey anomaly 151
Procollagenase 79
Prostacyclin synthase 81
Prostaglandin H synthase 114
Proteasome inhibitor 50

Proteinase 3, 23, 109, 180, 182
Protoporphyrin IX 33, 40, 116
Pseudomonas 41, 107, 111
Psoriasis 98

R569W mutant 48
Renal failure 150, 151
Retinoid acid response lement 95
Rheumatoid arthritis 73
Ricketiosis 42

Saccharomyces 41
Salicylhydroxamic acid 32
SARIp 11
Schistosoma 6
SeqA gene 55
SNAP 13
SNARE 13, 14
Soret band 40, 116
Staph.aureus 107, 188
Superoxid dismutase (SOD) 61, 62, 77, 162
Superoxide 102, 109
Surfactant protein A 81

Taurine 59, 78, 110
Technicon Hemalog Analyzer 136, 146, 157
Tetranitromethane 81
Thiocyanate 4, 52, 58, 129
Threonine 104
Thyroglobulin 129
Thyroid hormone response element 95
Thyroid peroxidase 114
–, biochemistry 129
–, gene 129
–, homology with other peroxidases 129
–, mutations 129
–, physiology 129
–, recombinant 131
thyroxine 130
TNF-alpha 109
Tobacco smoking 73
Trans Golgi network (TGN) 14, 29
Transplantation 73
Transport, vesicular 13
trichinella 6
Tryptophan 60, 64
Tryptophan radicals 72
Tubulin 80
Tubulin tyrosine ligase 70
tyrosine 6, 58, 59, 62, 64, 72, 74, 75, 78, 81, 102
tyrosine peroxide 6
Tyrosine radicals 72

Urate 59

Vasculitis 180, 186
Verdoperoxidase 2

Wegener granulomatosis 173, 177

Xanthin oxidase 54

Y173C mutant 49

Printing: Saladruck, Berlin
Binding: H. Stürtz AG, Würzburg